CHILDREN ARE FROM HEAVEN

CHILDREN
ARE
FROM
HEAVEN

Positive Parenting Skills for Raising Cooperative,
Confident, and Compassionate Children

JOHN GRAY, Ph.D.

Quill
An Imprint of HarperCollinsPublishers

A hardcover edition of this book was published in 1999 by HarperCollins
Publishers.

HarperCollins books may be purchased for educational, business, or sales
promotional use. For information please write: Special Markets Depart-
ment, HarperCollins Publishers Inc., 10 East 53rd Street, New York, NY
10022.

First Quill edition published 2001.

Designed by Nancy B. Field

Library of Congress Cataloging-in-Publication Data is available.

ISBN 0-06-093099-3

01 02 03 04 05 ❖/RRD 10 9 8 7 6 5 4 3 2 1

This book is dedicated with greatest love and affection to my wife, Bonnie Gray. I could not have written this book without her wisdom and insight. Her love, joy, and light have not only graced my life, but our children's as well.

Contents

Acknowledgments

I thank my wife, Bonnie, and our three daughters, Shannon, Juliet, and Lauren for their continuous love and support. Without their direct contributions, this book could not have been written.

I thank Diane Reverand at HarperCollins for her brilliant feedback and advice. I also thank Laura Leonard, my dream publicist, and Carl Raymond, Craig Herman, Matthew Guma, Mark Landau, Frank Fonchetta, Andrea Cerini, Kate Stark, Lucy Hood, Anne Gaudinier, and the other incredible staff at HarperCollins.

I thank my agent, Patti Breitman, for believing in my message and recognizing the value of *Men Are from Mars, Women Are from Venus* nine years ago. I thank my international agent, Linda Michaels, for getting my books published in more than fifty languages.

I thank my staff: Helen Drake, Bart and Merril Berens, Pollyanna Jacobs, Ian and Ellen Coren, Sandra Weinstein, Donna Doiron, Martin and Josie Brown, Bob Beaudry, Michael Najarian, Jim Puzan, and Ronda Coallier for their consistent support and hard work. I also thank Matt Jacobs, Sherri Rifkin, and Kevin Kraynick for their work in making marsvenus.com one of the best places on the Internet.

I thank my many friends and family members for their support and helpful suggestions: my brother, Robert Gray, my sister, Virginia Gray, Clifford McGuire, Jim Kennedy, Alan

Garber, Renee Swisco, Robert and Karen Josephson, and Rami El Batrawi.

I thank the hundreds of workshop facilitators who teach Mars–Venus workshops throughout the world and the thousands of individuals and couples who have participated in these workshops during the past fifteen years. I also thank the Mars–Venus counselors who continue to use these principles in their counseling practices.

I thank my dear friend, Kaleshwar, for his continued support and assistance.

I thank my mother and father, Virginia and David Gray, for all their love and support as they gently guided me to be the best parent I could be. And thanks to Lucile Brixey, who was like a second mother to guide me and love me.

I give thanks to God for the incredible energy, clarity, and support I received in bringing forth this book.

—*John Gray*
June 9, 1999

Introduction

After my first year of marriage, I was the father of a new baby and had two lovely stepdaughters. Lauren was the baby, Juliet was eight, and Shannon nearly twelve. Though my new wife Bonnie was a seasoned parent, this was my first experience. Having a baby, a child, and a preteen all at once was quite a challenge. I had taught many workshops with teens and children of all ages. I was very aware of the way children felt about their parents. I had also counseled thousands of adults, helping them resolve issues from their childhood. In areas where their parents' care was deficient, I taught adults how to heal their wounds by reparenting themselves. From this unique perspective, I began as a new parent.

At every step of the way, I would find myself automatically doing things my parents had done. Some things were good, others were less effective, and some were clearly not good at all. Based on my own experience of what didn't work for me and the thousands of people with whom I had worked, I was gradually able to find new ways of parenting that were more effective.

To this day, I can remember one of my first changes. Shannon and her mother, Bonnie, were arguing. I came downstairs to support Bonnie. At a certain point, I took over and yelled louder. Within a few minutes, I began to dominate the argument. Shannon became quiet, holding in her hurt and resentment. Suddenly, I could see how I was wounding my new stepdaughter.

In that moment, I realized that what I had done was a mistake. My behavior was not nurturing. I was behaving as my dad would when he didn't know what else to do. I was yelling and intimidating to regain control. Although I didn't know what else to do, I clearly knew that yelling and intimidating was not the answer. From that day on, I never again yelled at my kids. Eventually, my wife and I were able to develop other, more nurturing ways to regain control when our children misbehaved.

LOVE IS NOT ENOUGH

I am very thankful to my parents for their love and support, which helped me enormously, but, in many ways, in spite of the love, I was wounded by some of their mistakes. Healing those wounds has made me a better parent. I know they did their best with the limited knowledge they had regarding what children needed. When parents make mistakes in parenting, it is not because they don't love their children, but because they just don't know a better way.

The most important part of parenting is love and putting in time and energy to support your children. Although love is the most important requirement, it is not enough. Unless parents understand their children's unique needs, they are unable to give their children what children today need. Parents may be giving love, but not in ways that are most helpful to their child's development.

Without an understanding of their children's
needs, parents cannot effectively support
their children.

On the other hand, some parents are "willing" to spend more time with their children, but don't because they don't know what to do or their children reject their efforts. So many parents try to talk with their kids, but their kids just close up and say nothing. These parents are willing, but don't know how to get their kids to talk.

Some parents don't want to yell at, hit, or punish their children, but they just don't know another way. Since talking with their children has not worked, punishment or the threat of punishment is the only way they know.

To give up old ways of parenting, new ways must be employed.

Talking will work, but you have to learn first what children need. You have to learn how to listen so that children will want to talk to you. You have to learn how to ask so that children will want to cooperate. You have to learn how to give your children increasing freedom and yet maintain control. When a parent learns these skills, he or she can let go of outdated methods of parenting.

FINDING A BETTER WAY

As a counselor to thousands and teacher to hundreds of thousands, I was aware of what parenting behaviors didn't work, but I didn't yet know more effective solutions. To be a better parent, it was not enough just to stop doing things like punishing or yelling to control my children. To give up manipulating my children with the threat of punishment to maintain control, I had to find other equally effective methods. In developing the philosophy of *Children Are from Heaven* and the five skills of positive parenting, I gradually discovered an effective alternative to traditional parenting skills.

To be a better parent, it is not enough to stop doing things that don't work.

The skills of positive parenting contained in *Children Are from Heaven* took me more than thirty years to develop. For sixteen years as a counselor of adults with individual and relationship problems, I had a chance to study what didn't work in my clients' childhoods. Then, as a parent, during the next fourteen years I began to develop and use new and different parenting skills. These new insights and skills have not only worked in raising my own children, but also in thousands of other families.

Marge, a single parent, began using these skills with her oldest teenager daughter, Sarah, who wouldn't even talk with her and was on the verge of leaving home. When Marge shifted the way she communicated, they were able to resolve their issues. Sarah changed literally overnight. Before Marge took a *Children Are from Heaven* workshop, Sarah would scowl when her mother talked. Within a few months after the workshop, Sarah was talking about her life, listening, and cooperating with her mother.

Tim and Carol had difficulty with their youngest son, Kevin, who was three. He was always acting out, throwing tantrums, and controlling situations. By giving up spanking and using time outs instead, Kevin gradually had fewer tantrums. Tim and Carol learned how to regain control in their family by understanding how to nurture Kevin's unique needs.

Philip was a successful businessman. After taking a *Children Are from Heaven* workshop, he realized how much his children needed him, and what he could do to assist them in growing up. He had been raised mainly by his mother and didn't really know how much a father was needed. Once he learned what his children needed and what he could do, he was motivated to spend more time with his kids. He is grateful for this new information, not just because his children are happier, but because he is happier. He was missing out on the joys of parenthood and he didn't even know it.

Many men who are not involved in parenting don't realize the joys they are missing.

Tom and Karen were always fighting about how to raise their children. Since they were raised differently, they would argue about how to discipline or raise their children. After taking a *Children Are from Heaven* workshop, they had a common approach to raising their kids. The children not only benefited from more effective support, but also because their parents stopped fighting all the time.

There are endless stories of families who have benefited from the new insights and skills of *Children Are from Heaven*. If you have any doubts regarding their validity, just try them and see the results. The effectiveness of these skills is easy to prove. As you begin to use them, they work immediately.

The effectiveness of these skills is easy to prove. Use them; they work immediately.

Each suggestion in *Children Are from Heaven* simply makes sense. In many cases, your experience of reading

Children Are from Heaven will clarify what you already felt was true or right for you. In other cases, these new insights will point out where you have made some mistakes and answer many of your questions. Although *Children Are from Heaven* does not deal with every problem you will encounter, it provides a whole new approach for problem solving. You still solve the problems, but with a different and more effective approach. This new way of understanding children will assist you in coming up with your own unique day-to-day solutions.

Children Are from Heaven is a broad practical philosophy of parenting that works at every age. The new insights and skills work for infants, toddlers, young children, preteens, and teens. Even if your teens were not raised with these skills, they will quickly begin to respond to them.

Children Are from Heaven is a broad
practical philosophy of parenting that works
at every age.

In my own experience, I found that my two stepdaughters responded immediately to this new nonpunishing approach. Even though they had been raised with some of the old methods, like punishment or yelling, the new approach was effective. Children at any age, regardless of their past, begin to cooperate more as a result of using these new skills.

These techniques work even when children have been raised with neglect, abuse, or cruel punishment. Certainly, neglected or abused children may have unique behavioral problems, but these are more effectively corrected or solved as soon as this new approach is employed. Children are incredibly resilient and adaptable when given the right kind of loving support.

THE NEW CRISIS OF PARENTING

The Western free world is experiencing a crisis in parenting. Every day, there are increasing reports of child and teen violence, low self-esteem, Attention Deficit Disorder, drug use, teen pregnancy, and suicide. Almost all parents today are questioning both the new and old ways of parenting. Nothing seems to be working, and our children's problems continue to increase.

Some parents believe that these problems come from being too permissive and giving children too much, while others contend that outdated practices of parenting, like spanking and yelling, are responsible. Others believe these new problems are caused by negative changes in society.

Too much TV, advertising, or too much violence and sex on TV and in movies are pegged by many as the culprits. Certainly society and how it influences our children is part of the problem, and some helpful solutions can be legislated by the government, but the biggest part of the problem starts at home. Our children's problems begin in the home and can be solved at home. Besides looking to change society, parents must also realize that they hold the power to raise strong, confident, cooperative, and compassionate children.

Our children's problems begin in the home,
and can be solved at home.

To cope with changes in society, parents need to change their parenting approach. During the past two hundred

xxiv   Introduction

years, society has made an historic and dramatic change toward greater individual freedom and rights. Even though our modern Western society is now organized by the principles of freedom and human rights, parents still use parenting skills from the Dark Ages.

Parents need to update their parenting skills to raise healthy and cooperative children and teens. Businesses know that if they are to stay competitive in the free market, they need to keep changing and updating. Likewise, if parents want their children to be able to compete in the free world, they must prepare their children with the most effective and modern approaches to parenting.

LOVE- VERSUS FEAR-BASED PARENTING

In the past, children where controlled by dominance, fear, and guilt. To motivate good behavior, children were made to believe they were bad and unworthy of good treatment if they were not obedient. The fear of losing love and privileges was a strong deterrent. When this didn't work, stronger punishment was given to generate even more fear and to break the will of a child. An unruly child was often called strong-willed. Ironically, from the perspective of positive parenting, nurturing a strong will is the basis of creating confidence, cooperation, and compassion in children.

Nurturing and not breaking a child's will is
the basis of creating confidence, cooperation,
and compassion in children.

Past parenting approaches sought to create obedient children. The goal of positive parenting is to create strong-willed but cooperative children. A child's will doesn't have to be broken in order to create cooperation. Children are from heaven. When their hearts are open and their will is nurtured, they actually are more willing to cooperate.

The goal of positive parenting is to create willful but cooperative children.

Past parenting approaches were aimed at creating good children. Positive parenting creates compassionate children, who don't have to be threatened to follow rules, but spontaneously act and make decisions from an open heart. They do not lie or cheat because it is against the rules, but they are fair and just. Morality is not imposed on these children from outside, but emerges from within and is learned by cooperating with their parents.

Rather than seeking to create good children, positive parenting seeks to create compassionate children.

Past parenting approaches focused on creating submission; positive parenting aims to develop confident leaders, who are capable of creating their own destiny, not just passively following in the footsteps of others before them. These confident children are aware of who they are and what they want to accomplish.

Confident children are not easily swayed by peer pressure nor do they feel the need to rebel.

These strong children are not easily swayed by peer pressure nor do they feel a need to rebel in order to be themselves. They think for themselves, yet remain open to the assistance and help of their parents. As adults, they are not held back by the limited beliefs of others. They follow an inner compass and make decisions for themselves.

CHILDREN TODAY ARE DIFFERENT

Just as the world today is different, our children are different. They no longer respond to fear-based parenting. The old fear-based approaches actually weaken a parent's control. The threat of punishment only turns children against their parents and causes them to rebel. The intimidation of yelling and spanking no longer creates control, but simply numbs a child's willingness to listen and cooperate. Parents are seeking better communication with their children to prepare them for the increased pressures of life today but, unfortunately, they are still using outdated approaches for parenting.

The threat of punishment only turns children
against their parents and causes them to rebel.

I remember my dad making this mistake. He would try to control his six boys and one daughter with threats of punishment. He had been a sergeant in the military, and this was the only way he knew. In some ways, he treated us like army privates. Whenever we would resist his control, he would regain control with the threat of punishment. Though this parenting style worked to some degree in his generation, it didn't work for mine, and it clearly is not working for our children today.

When his threat didn't result in obedience, my father

would increase the threat. He would say, "If you keep talking to me like that, you are grounded for a week."

When I continued to resist, he would say, "If you don't stop, it will be two weeks."

When I persisted, he would say, "Okay then, you are grounded for one month, now go to your room."

Upping the punishment has no real positive effect and only engenders greater resentment. For the whole month, I just reflected on how unfair he was. Instead of increasing my willingness to cooperate, his action pushed me farther away. He would have had a much more positive influence if he had just said, "Since you are not respecting what I am saying, I want you to take a time out for ten minutes."

Punishment in the past was used to break a strong-willed child. Although it may have worked to create obedience, it doesn't work today. Children are now more sophisticated and aware. They recognize what is unfair and abusive and will not tolerate it. They will resent and rebel. Most importantly, punishment and the threat of punishment break down the lines of communication. Instead of being a part of the solution, you the parent become a part of the problem.

Punishment makes you, the parent, an enemy
to hide from instead of a parent to turn to
for support.

When parents yell at children, it just numbs their ability to hear. To succeed in school and, more importantly, to compete in the free market or experience success in a lasting relationship, adults today need better communication skills. These skills are most effectively learned when children listen to their parents and parents listen to their children.

Children listen to their parents when parents
learn how to listen to their children.

What happens when you listen to music at loud levels? You lose your hearing. The same thing happens when parents yell or make demands all the time. When parents today yell or communicate the way their parents did, it has a different effect. Children today will just be turned off, and parents will lose control.

GIVING UP PUNISHMENT

In previous generations, societies were suppressed, controlled, and manipulated by strong, punishing dictators, but it is not so today. People will not stand for injustice and the violation of human rights; they will revolt instead. People have sacrificed their lives for the principles of democracy.

In a similar way, children today will not accept the threat of punishment. They will revolt. Children today feel more intensely the injustice of punishment. When punishment goes in, it comes back out as increased resistance, resentment, rejection, and rebellion. Children today are rejecting their parents' values and rebelling against parental control at younger and younger ages.

Before they are psychologically mature or prepared to let go of their parents' support, children and teens are pulling away and rejecting the support that is so important for their development. They long to be free of their parents' control at a time when they need that control to develop in a healthy manner.

> Before they are psychologically prepared,
> children and teens are rejecting necessary
> parental support.

Many parents recognize that the old methods of punishment don't work, but they just don't know another way. They hold back from punishing, but that doesn't work either. Permissive parenting doesn't give children the parental control they need. When given an inch of power, these children take a mile. Children quickly learn to use their freedom to manipulate and control parents.

When children are allowed to use strong, negative moods, feelings, and tantrums to get their way, they are in control. When a child is in control, they are out of their parents' control. In many ways, they will develop some of the same problems of children who are raised with outdated fear-based skills.

> When children are in control, they are out of
> their parents' control.

Whether a child is raised with fear-based skills or permissive skills, if the child doesn't experience that his parents are in control, he will rebel or reject any attempts a parent makes to regain or maintain control. Disconnected from his parents' support, his development will be restricted. By using the skills of positive parenting in *Children Are from Heaven,* parents can give their children the freedom and leadership they need to develop a strong and healthy sense of self.

THE RESULTS OF FEAR-BASED PARENTING

The old fear-based practices of managing our children through intimidation, criticism, disapproval, and punishment have not only lost their power but are counterproductive. Children are more sensitive than in previous generations. They are capable of much more, but are also influenced in a negative way by old parenting skills like yelling, spanking, hitting, punishing, grounding, disapproving, humiliating, and shaming. When children were more thick-skinned, these approaches were useful, but today they are outdated and counterproductive.

In the past, punishing children by spanking made them fear authority and follow the rules. Today it has the opposite effect. Violence in means violence out. This is a symptom of being more sensitive. Children today can be more creative and intelligent than in previous generations, but they are also more influenced by outer conditions.

When children are more sensitive, violence in
means violence out.

Children today can best learn to respect others, not by fear tactics, but through imitation. Children are programmed to imitate their parents. Their minds are always taking pictures and making recordings to mimic and follow whatever you say or do. They practically learn everything through imitation and cooperation.

When parents model respectful behavior, children gradually learn how to respect others. When parents learn how to remain cool, calm, and loving while dealing with a child throwing a tantrum, that child gradually learns how to remain cool, calm, and loving when strong feelings come up.

Parents can stay calm, cool, loving, and respectful when they learn what to do when children go out of control.

Parents can stay calm and cool when
they learn what to do when children go
out of control.

If you hit children to regain control, children learn that aggression is the answer when they feel out of control. Many times I have witnessed a mother hitting her son, saying, "Stop hitting your brother." She wants him to understand how it feels, but hitting is not the answer. By hitting her son, she reinforces his tendency to hit or use aggression.

Later on, when he is not getting what he wants, he will automatically resort to acting out his anger by either direct or passive aggression. Although spanking or hitting children worked in the past, it backfires today. Fear-based parenting methods restrict our children's natural development and make our job as parents less fulfilling and more time consuming.

NOT ENOUGH TIME TO PARENT

Parents today have less time than ever to devote to parenting. For this reason, it is essential that they learn what is most important for their children. This knowledge not only helps them to use their time more efficiently, but also motivates them to create more time. A greater awareness of their children's needs naturally motivates parents to spend more time with their children.

In dealing with stress and pressure, many adults often devote time to what they feel they have to do and can do. Women commonly feel overwhelmed with all the things they have to do. Men feel primarily focused on what they can do. When fathers don't know what they can do to help their children, they often do nothing. When mothers are not aware of what their children need, they often make others things more important.

When parents learn what their children really need, they are less motivated to create money to acquire things and more motivated to create time to enjoy their family. The greatest wealth for a parent today is time. Parents begin to find more time to be with their children when they recognize what they have to do and can do.

UPDATING YOUR PARENTING SKILLS

By reading *Children Are from Heaven,* you will learn practical ways to update your parenting skills. You will not only learn what doesn't work, but what you can do instead. You will learn new ways of motivating your children to cooperate and excel without having to use fear tactics.

Children today do not need to be motivated by the fear of punishment. They have the innate ability to know what is right and wrong when given an opportunity to develop this ability. Instead of being motivated by punishment or intimidation, they can be easily motivated by reward and the natural, healthy desire to please their parents.

In the first eight chapters of *Children Are from Heaven,* you will learn to use the different skills of positive parenting to improve communication, increase cooperation, and motivate your children to be all they can be. In last six chapters, you will learn how to communicate the five most important messages your children need to hear again and again.

The five positive messages are:

1. It's okay to be different.

2. It's okay to make mistakes.

3. It's okay to express negative emotions.

4. It's okay to want more.

5. It's okay to say no, but remember mom and dad are the bosses.

These five messages will set your children free to develop their God-given abilities. When practiced correctly with the different skills of positive parenting, your child will develop the necessary skills for successful living. Some of these skills are: forgiveness of others and themselves, sharing, delayed gratification, self-esteem, patience, persistence, respect for others and themselves, cooperation, compassion, confidence, and the ability to be happy. With this new approach, along with your love and support, your children will have the opportunity to develop fully during each stage of their growth.

With these new insights, you will have the confidence needed to raise your children well and to sleep soundly at night. When questions and confusion arise, you will have a powerful resource to return to again and again to give you support and to remind you of what your children need and what you can do for them.

Most of all, you remember that children *are* from heaven. They already have within them what they need to grow. Your job as parent is only to support their process of growth. By applying the five messages and positive-parenting skills, you will not only enjoy the confidence that you doing exactly what is needed, but know that, with your help, your children will be able to create the life they were meant to live.

1

Children Are
from Heaven

All children are born innocent and good. In this sense our children are from heaven. Each and every child is already unique and special. They enter this world with their own particular destiny. An apple seed naturally becomes an apple tree. It cannot produce pears or oranges. As parents, our most important role is to recognize, honor, and then nurture our child's natural and unique growth process. We are not required in any way to mold them into who we think they should be. Yet we are responsible to support them wisely in ways that draw out their individual gifts and strengths.

Our children do not need us to fix them or make them better, but they are dependent on our support to grow. We provide the fertile ground for their seeds of greatness to sprout. They have the power to do the rest. Within an apple seed is the perfect blueprint for its growth and development. Likewise, within the developing mind, heart, and body of every child is the perfect blueprint for that child's development. Instead of thinking that we must do something to make our children good, we must recognize that our children are already good.

Within the developing mind, heart, and body
of every child is the perfect blueprint for that
child's development.

As parents we must remember that Mother Nature is always responsible for our children's growth and development. Once, when I asked my mother the secret of her parenting approach, she responded this way: "While raising six boys and one girl, I eventually discovered there was little that I could do to alter them. I realized it was all in God's hands. I did my best and God did the rest." This realization allowed her to trust the natural growth process. It not only made the process easier for her, but also helped her to not get in the way. This insight is important for every parent. If one doesn't believe in God, one can just substitute "genes"—It's all in the genes.

By applying positive-parenting skills, parents can learn to support their children's natural growth process and to avoid interfering. Without an understanding of how children naturally develop, parents commonly experience unnecessary frustration, disappointment, worry, and guilt and unknowingly block or inhibit parts of their children's development. For example, when a parent doesn't understand a child's unique sensitivity, not only is the parent more frustrated, but the child gets the message something is wrong with him. This mistaken belief, "something is wrong with me," becomes imprinted in the child and the gifts that come from increased sensitivity are restricted.

EVERY CHILD HAS HIS OR HER OWN UNIQUE PROBLEMS

Besides being born innocent and good, every child comes into this world with his or her own unique problems. As parents, our role is to help children face their unique challenges. I grew up in a family of seven children and, although we had the same parents and the same opportunities, all seven children turned out completely different. I now have three daughters ages twenty-five, twenty-two, and thirteen. Each one is, and has always been, completely different, with a different set of strengths and weaknesses.

As parents, we can help our children, but we cannot take away their unique problems and challenges. With this insight, we can worry less, instead of focusing on changing them or solving their problems. Trusting more helps the parent as well as the child. We can let our children be themselves and focus more on helping them grow in reaction to life's challenges. When parents respond to their children from a more relaxed and trusting place, children have a greater opportunity to trust in themselves, their parents, and the unknown future.

Each child has his or her own personal destiny. Accepting this reality reassures parents and helps them to relax and not take responsibility for every problem a child has. Too much time and energy is wasted trying to figure out what we could have done wrong or what our children should have done instead of accepting that all children have issues, problems, and challenges. Our job as parents is to help our children face and cope with them successfully. Always remember that our children have their own set of challenges and gifts, and there is nothing we can do to alter who they are. Yet we can make sure that we give them the opportunities to become the best they can be.

Children have their own set of challenges and gifts, and there is nothing we can do to alter who they are.

At difficult times, when we begin to think something is wrong with our children, we must come back to remembering that they are from heaven. They are perfect the way they are and have their own unique challenges in life. They not only need our compassion and help, but they also need their challenges. Their unique obstacles to overcome are actually necessary for them to become all that they can become. The problems they face will assist them in finding the support they need and in developing their special character.

Children need compassion and help, but they also need their unique challenges to grow.

For every child, the healthy process of growing up means there will be challenging times. By learning to accept and embrace the limitations imposed by their parents and the world, children can learn such essential life skills as forgiveness, delayed gratification, acceptance, cooperation, creativity, compassion, courage, persistence, self-correction, self-esteem, self-sufficiency, and self-direction. For example:

- Children cannot learn to be forgiving unless there is someone to forgive.

- Children cannot develop patience or learn to delay gratification if everything comes their way when they want it.

- Children cannot learn to accept their own imperfections if everyone around them is perfect.

- Children cannot learn to cooperate if everything always goes their way.

- Children cannot learn to be creative if everything is done for them.

- Children cannot learn compassion and respect unless they also feel pain and loss.

- Children cannot learn courage and optimism unless they are faced with adversity.

- Children cannot develop persistence and strength if everything is easy.

- Children cannot learn to self-correct unless they experience difficulty, failure, or mistakes.

- Children cannot feel self-esteem or healthy pride unless they overcome obstacles to achieve something.

- Children cannot develop self-sufficiency unless they experience exclusion or rejection.

- Children cannot be self-directed unless they have opportunities to resist authority and/or not get what they want.

In a variety of ways, challenge and growing pains are not only inevitable, but also necessary. As parents, our job is not to protect our children from life's challenges but to help them successfully overcome them and grow. Throughout *Children Are from Heaven* you will learn new positive parenting skills to assist your children in responding to life's challenges and setbacks. If you are always solving their problems, they do not find within themselves their innate abilities and skills.

Life's obstacles can occur to strengthen your children in unique ways and draw out the best in them. When a butterfly emerges from its cocoon, there is a great struggle. If you were to cut open the cocoon in order to spare the new butterfly this struggle, it would soon die. The struggle to get out is needed to build the wing muscles. Without the struggle, the butterfly will never fly, but will die instead. In a similar way, for our children to grow strong and fly free in this world, they need particular kinds of struggle and a particular kind of support.

To overcome their unique challenges, every child needs a particular kind of love and support. Without this support, their problems will become magnified and distorted, sometimes to the point of mental disease and criminal behavior. Our job as parents is to support our children in special ways so that our children become stronger and healthier. If we interfere and make it too easy, we weaken children, but, if we make it too tough and don't help enough, then we deprive them of what they need to grow. Children cannot do it alone. A child cannot grow up and develop all the skills for successful living without the help of their parents.

THE FIVE MESSAGES OF POSITIVE PARENTING

There are five important positive messages to help your children find within themselves the power to meet life's challenges and develop their full inner potential. Throughout *Children Are from Heaven*, we will explore a variety of new parenting skills based on communicating each of these five messages. The five messages are:

1. It's okay to be different.

2. It's okay to make mistakes.

3. It's okay to express negative emotions.

4. It's okay to want more.

5. It's okay to say no, but remember mom and dad are the bosses.

Let's explore each of these messages in greater detail.

1. It's okay to be different. All children are unique. They have their own special gifts, challenges, and needs. As parents, our job is to be able to recognize what their special needs are and to nurture them. Boys in general will have special needs that are not as important for girls. Likewise, girls will have needs that may not be that important for boys. In addition, every child regardless of gender has special needs associated with his or her particular challenges and gifts.

Children are also different in the way they learn. It is essential for parents to understand this difference, otherwise they may begin comparing children and become unnecessarily frustrated. When it comes to learning a task, there are three kinds of children: runners, walkers, and jumpers. Runners learn very quickly. Walkers learn in a steady manner and give clear feedback that they are making progress. Finally, there are the jumpers. Jumpers tend to be more difficult to raise. They don't seem to be learning anything or making any progress, and then one day they make one jump and have it. Jumpers are like late bloomers. Learning takes more time for them.

Parents learn the importance of expressing love in gender-specific ways. For example, girls often need more caring, but too much caring can make a boy feel as if you don't trust him. Boys need more trust, though too much trust for a girl may be interpreted as not caring enough. Fathers mistakenly tend to give their daughters what boys need, while mothers mistakenly tend to give boys the support girls would need. Understanding how boys and girls have different needs helps parents be more successful in nurturing their children. In addition, mothers and fathers argue less about each other's parenting style. Daddies Are from Mars and Mommies Are from Venus.

2. It's okay to make mistakes. All children make mistakes. It is perfectly normal and to be expected. Making a mistake does not mean something is wrong with you, unless your parents react as if you should not have made a mistake. Mistakes are natural, normal, and to be expected. The way children learn this is primarily by example. Parents can most effectively teach this principle by making sure they acknowledge their own mistakes in dealing with and supporting their children and each other.

When children see their parents apologizing on a regular basis, they gradually learn to be accountable for their own mistakes. Instead of teaching children to apologize, parents demonstrate. Children learn from role models not by lectures. Not only do children learn to be more responsible, but, by repeatedly forgiving their parents for their mistakes, children gradually learn the important skill of forgiveness.

Children come into this world with the ability to love their parents, but they cannot love or forgive themselves. They learn to love themselves by the way they are treated by their parents and how their parents react when they make mistakes. When children are not shamed or punished for their mistakes, they have a better chance to learn the most important skill: the ability to love themselves and accept their imperfections.

This skill is learned by repeatedly experiencing that their parents make mistakes and are still lovable. Shaming or punishing prevents children from developing self-love or the ability to forgive themselves. Throughout *Children Are from Heaven*, parents learn effective alternatives to spanking, shaming, and punishing that involve new ways of asking instead of ordering, giving rewards instead of punishments, giving time outs instead of spanking. These new positive parenting skills are described in greater detail through chapters 3 through 8. A time out, if given correctly and persistently, is just as powerful a deterrent as spanking and punishment.

3. It's okay to have negative emotions. Such negative emotions as anger, sadness, fear, sorrow, frustration, disappointment, worry, embarrassment, jealousy, hurt, insecurity, and shame are not only natural and normal, but an important part of growing up. Negative emotions are always okay and they need to be communicated.

Parents must learn to create appropriate opportunities for children to feel and express their negative emotions. Although negative emotions are always okay, how, when, and where our children express them is not always appropriate. Tantrums are an important part of a children's development, but they need to learn the time and place. On the other hand, you must make sure that you are not placating a child to avoid a tantrum, otherwise tantrums will come out when you don't have an opportunity to give your child a time out and deal with the problem at hand more effectively.

New communication skills must be learned and practiced to increase children's awareness of what they are feeling, otherwise they will go out of control, resist your authority, and act out on pent-up feelings. In this book, parents will learn to deal effectively with their own upset feelings. What parents suppress, their children will express in addition to their own upset feelings. This principle explains why children lose control at the most inconvenient times, particularly at stressful and overwhelming times when we are trying to keep a lid on our own feelings.

Positive parenting involves not making children responsible for how parents feel. When children get the message that their feelings and the needs for understanding and affection underlying those feelings are an inconvenience, they will begin to suppress their feelings and disconnect from their true self and all the gifts that come from being authentic.

"Enlightened" parents, who recognize the importance of feelings, often make the mistake of teaching their children to feel by sharing their own emotions too much. The best way to teach awareness of feelings is to listen and help identify feelings with empathy. Parents can best share their own negative feelings by telling stories of how they felt growing up in reaction to some of their challenges in life. The downside of sharing your own negative feelings with your children is that it makes children overly responsible. These children assume too much blame and disconnect from their own inner feelings. They eventually pull away and stop talking to you.

For example, telling a child, "When you climb that tree, I am afraid you will fall" has the gradual effect of making the child feel manipulated and controlled by negative feelings. Instead, an adult should say, "Climbing trees is not completely safe. I only want you to climb when I am around." This is not only more effective, but it also teaches children not to make decisions based on negative emotions. The child cooperates not to protect the parent from the discomfort of feeling afraid, but because the parent has asked them to do something.

Parents can help their children develop an increased awareness of feelings by empathizing, acknowledging, and listening, not by sharing their own feelings. Sometimes, even directly asking children how they feel or what they want may give them too much power. New listening skills must be used to draw out feelings and to understand a young child's want and needs. "Permissive" parents will learn how not to be ruled by or manipulated by children's wants and feelings. "Demanding" parents will learn the many ways they unknowingly shame their children for having negative feelings.

By learning to feel and communicate negative emotions, children most effectively learn to individuate from their parents, developing a strong sense of self, and gradually discover within themselves a wealth of inner creativity, intuition, love, direction, confidence, joy, compassion, conscience, and the ability to self-correct after making a mistake. All these advanced life skills, which make a person shine out in this world and achieve great success and fulfillment, come from staying in touch with feelings and being able to let go of negative feelings. Successful people feel their losses, but they bounce back because they have the ability to let go of negative feelings. Most people who do not achieve personal success are either numb to their inner feelings, make decisions based on negative feelings, or just remain stuck in negative feelings and attitudes. In each case, they are held back from making their dreams come true.

4. It's okay to want more. Too often children get the message that they are wrong, selfish, or spoiled for wanting more or for getting upset when they don't get what they want. Parents are too quick to teach the virtues of gratitude instead of giving their children permission to want more. "Be grateful for what you have" is too quick a reply to a child's desire for more.

Children don't know how much is acceptable to ask for and should never be expected to know. Even as adults, we still have difficulty determining how much we can ask for without offending or appearing too demanding or ungrateful. If adults have difficulty, then clearly we should not expect our children not to.

Positive parenting skills teach children how to ask for

what they want in ways that are respectful to others. At the same time, parents will learn how to say no without getting upset. Children will feel free to ask for what they want, knowing that they will not be shamed. They will also recognize that just because they ask doesn't mean they will get what they want.

Unless they are free to ask for what they want, children never clearly learn what they can get and what they can't. In addition, by asking for what they want, they quickly develop incredible negotiating skills. Most adults are very poor negotiators. They don't ask unless they expect a yes. If they get a no, they usually just accept it and walk away either submissive, secretly resentful, or outwardly angry.

When given the freedom to ask for what they want, children's inner power to get what they want has a chance to blossom. As adults they will not take no for an answer. As children, they learn to negotiate and will often motivate you to give them what they want. There is big difference between being manipulated by a whiny child and being motivated by a brilliant negotiator. Positive parents do maintain control throughout every negotiation and clearly set limits on how long it can go on.

By giving your child permission to ask for more, you give that child the gift of direction purpose, and power in life. Too many women today feel powerless, because they were never given permission to ask for more. They were taught to care more about what others needed and shamed for getting upset when they didn't get what they wanted or needed.

One the most important skills a father or mother can teach a girl is how to ask for more. Most women did not learn this lesson as children. Instead of asking for more, they indirectly ask for more by giving more and hoping someone will give back to them without their having to ask. This inability to ask directly prevents them from getting what they want in life and in their relationships.

While girls need permission to want more, boys need a particular kind of support when they don't get more. Quite often a boy will set his goals really high, and parents will try to talk him out of his goals, because they want to protect him from being disappointed. They do not realize that more important than achieving goals is being able to cope with disappointment so that he can rise again to move toward his goals. Just as girls need a lot of support in asking for what they want, boys need extra support to identify their feelings and move through them. For boys, this is best accomplished by asking for details of what happened while being *extremely* careful not to offer any advice or "help." Even too much empathy can turn him off to talking about what happened.

Mothers often make the mistake of asking too many questions. When pushed to talk, many boys stop. When given suggestions on how to cope, boys particularly will back off. At a time when he already feels beaten, he doesn't need someone to make him feel worse by telling him how to solve the problem or what he did to contribute to the problem.

For example, he feels disappointed that he didn't score well on a test and his mother says, in a caring way, "I think that if you would have watched less TV and taken more time to study then you would have done better. You are really smart, you are just not giving yourself a chance." Clearly she thinks she is being loving, but in this context it is clear why he would stop revealing to her what is bothering him. She has offered unsolicited advice, and he feels both criticized and not trusted to solve his problem.

5. It's okay to say no, but remember mom and dad are the bosses. Children need permission to say no, but, just as important, they need to know that their parents are in charge. Besides giving children permission to want more and to negotiate, the permission to say no really gives them power. Most parents are afraid of giving children that much power because they may easily become spoiled. One of the biggest problems today with children is that they have been given too much freedom. Parents have sensed that their children deserve more power, but they have not learned how to remain the boss. Unless they employ other positive parenting techniques like consistent time outs to maintain cooperation, their children become too demanding, selfish, and irritable. When parents remain in control, it is then effective to give their children more power.

Letting children say no opens the door for them to express feelings and to discover what they want and then negotiate. It does not mean you will always do what the child wants. Even though children can say no, it doesn't mean they will get their way. What children feel and want will be heard and this in itself often makes them much more cooperative. More importantly, it allows children to be cooperative without having to suppress their true self.

There is a big difference between adjusting your wants and denying your wants. Adjusting your wants means shifting what you want to what your parents want. Denying means suppressing your wants and feelings and submitting to your parents wants. Submission results in a breaking of the child's will. After a horse is broken, it becomes submissive and thus cooperative, but it also loses a big part of its free spirit.

Analysis of parenting practices in pre-Nazi Germany revealed that children were severely shamed and punished for resisting authority. They had no permission to resist or say no. In retrospect, we can see clearly on a much bigger scale how breaking the will of your children can make them mindless and heartless followers of strong but maniacal authoritarian leaders. When a person does not have a strong sense of self, he is easy prey for others to manipulate and abuse. Without a strong sense of self, a person will even be attracted to abusive relationships and situations, because of feelings of unworthiness and fear of asserting his own will.

Adjusting one's will and wish is called *cooperation*, submitting one's will and wish is *obedience*. Positive parenting practices seek to create cooperative children not obedient children. It is not healthy for children to follow their parents' will mindlessly or heartlessly. Giving children permission to feel and verbalize their resistance when it occurs not only helps children develop a sense of self, but also makes children more cooperative. Obedient children just follow orders; they do not think, feel, or contribute to the process. Cooperative children bring their full self to every interaction and thus are able to thrive.

Positive parenting practices seek to create
cooperative children, not obedient children.

Cooperative children may still want what they want, but what they want most is to please their parents. Giving children permission to say no does not mean giving them more control; it actually gives the parent more control. Each time children resist and the parents maintain control, the children are able to experience that mom and dad are the bosses. This is the main reason that giving children a time out is so valuable.

When children are misbehaving or not cooperating, they

are simply out of control. They are out of your control. They are not in cooperation with your will and wish. To restore cooperation, a parent needs to regain control of them through picking them up and moving them into a time out. God makes children little so that we can pick them up and move them.

In a time out children have the freedom to resist and express all their feelings, but they are still restricted to a time out for a set time. Generally speaking, all that a child needs is one minute for every year of his or her life. A four year old only needs four minutes. The containment of a time out is all that is required for children to feel once again the security of being under your control and connected to you as the boss. Automatically, the negative feelings lift off, and the child reconnects to the healthy desire to please and cooperate.

Parents who are too permissive or don't give their children enough time outs unknowingly make their children more insecure. The child begins to feel they have the power to control and, because they are not ready to be in charge (although they like the power), they feel insecure. Imagine being given the responsibility to hire two hundred workers to build a building in six months. Or, imagine that you were handed a bleeding person recently shot with a gun and asked to operate on him and to remove the bullet. If you were not trained for either of these jobs, you would suddenly feel very insecure. When children begin to feel the thrill of being the boss, they also begin to feel very insecure and demanding.

A demanding or "spoiled" child generally needs more time outs. A spoiled teenager may need more than time out in his or her room. In some cases, time spent with supervision in a developing country, or in the woods with a guide, or staying with a favorite aunt, uncle, or grandparent will help teenagers regain their true self and their need for someone else to be boss. By feeling out of control and depending on someone else, a teenager is humbled. They can come back to feeling their need for parents and the desire to please them.

To be secure, children should feel heard, but always know that they are not the boss.

Children are basically programmed to one prime directive. Deep inside they only want to please their parents. Positive parenting communication skills strengthen this prime directive so that children are more willing to follow a parent's will and wish. To balance this yielding tendency, children need permission to resist and say no. This resistance allows them to develop a healthy sense of self.

Children who don't get this opportunity go through unnecessary rebellion around puberty. Although a teenager still needs guidance in life, they feel huge urges just to do the opposite of whatever is your will and wish, if they have not developed a sense of self.

Many parents take it for granted that their children need to pull away from them at this time and rebellion is perfectly normal. Rebellion is only a normal reaction for children who did not get the support needed at an earlier stage. When children experience the permission to say no, but then cooperate with their parents, they have a healthy sense of self and don't need to rebel at puberty. They still pull away, but they don't rebel and they keep coming back for love and support.

Positive parenting also explores ways of improving communication with teenagers, who were not raised with these five positive parenting messages. It is never too late to be a great parent and inspire cooperation from your children. No matter when you start, by applying the five messages of positive parenting, you will hold the power to improve communication, create cooperation, and draw from your children the best they can be.

A VISION OF POSSIBILITIES

Even with a greater understanding of the five messages of positive parenting, being a good parent is not easy. It is a learn-as-you-go process. Parenting pushes you beyond your limits of how much you thought you could give. Yet, no matter how good you get, you always find yourself once again in uncharted territory wondering: "What do I do now?" A clear vision of possibilities is needed. Fortunately, you can return to this guide again and again. When something doesn't seem to be working, or you don't know what to do, review the different messages of positive parenting. You will discover what is missing and be better equipped to do the right thing.

As parents, we don't get a lot of practice to prepare ourselves or to perfect our parenting abilities. Suddenly we are faced with the awesome responsibility of caring for a vulnerable child, and we are not always certain what is best for them. Even though we remind ourselves that children are from heaven and that they have their own unique potential destiny, their future is literally in our hands. How we hold and care for them greatly influences their ability to succeed in manifesting their full potential.

Parenting requires a tremendous commitment on our part but our children are certainly worth it. Parents only "back off" or withdraw from parenting when they don't know what to do or when what they do seems to make matters worse. Studying the easy-to-understand (but not always easy-to-remember) principles of positive parenting will always remind you that you are needed and that by making a few adjustments you can succeed in giving your children what they need.

Always remember that no one can do it better than you can. Although your children come from heaven, they also come from you and they need you. Learning how to parent is the most worthwhile study a person can make if planning to have a family. Without the understanding of positive parenting, most parents have no idea how important they are to their children and their children's future. Not only do their children miss out, but they do as well.

Parenting is a difficult job, but it is also the most rewarding. To be a parent is an awesome responsibility and a great honor. Now, with an awareness of what our children really need from us, parents can fully understand how much their help is needed. This clear insight into our responsibility allows us to feel the true dignity of being a parent and to take pride in doing what is required in caring for our family.

By fully committing yourself to the new principles of positive parenting, you are a courageous pioneer exploring new territory, a brave hero creating a new world and, most important, you are giving your children the opportunities for greatness that you never had.

Even with this guide by your side, you will still make mistakes, but then you will be able to use your mistakes to teach your children the important skill of forgiveness. We can't always give our children what they need or want, but we can help them respond to their disappointments in healthy ways that make them stronger and more confident. You will still be unable to always be there when they need you, but you will know how to react to their feelings and unmet needs in a way that heals their emotional wounds and makes them feel loved and supported once again. Using the five messages of positive parenting and remembering that children are from heaven, will help you give your children the best preparation they could have to make all their dreams come true, which is what all parents want for their children.

2
What Makes the Five Messages Work

To apply the five messages of positive parenting, we first have to understand the right conditions for them to work. These new parenting skills will not work if we keep control of our children with threats of spanking, punishment, or guilt. Fear-based parenting numbs our children's ability to respond to positive parenting. On the other hand, if we don't know how to replace spanking and punishment with more positive ways to create cooperation, the five messages will not work as well. It is not enough just to stop punishing our children; we must apply new skills to create cooperation, motivation, and control.

If parenting is based on fear, children will not respond to the five messages. For this new approach to work, parents must let go of outdated fear-based practices of parenting. To flip back and forth doesn't work. You can't treat children as if they are good and innocent in order to draw out their inner greatness, and then spank them for being bad a week later.

It doesn't work to treat children as if they are good and innocent, and then spank them for being bad a week later.

If we want our children to feel good about themselves, we have to stop making them feel bad. If we want our children to feel confident, we have to stop controlling them with fear. If we want our children to respect others, we must learn how to show them the respect they deserve. Children learn by example. If you manage them with violence, they will resort to violence, or at least sometimes cruel or insensitive behavior, when they don't know what to do.

THE PRESSURE OF PARENTING

Because of the invention of Western psychology, we are now much more aware of the profound influence childhood has on our success later in life. Both our ability to create outer success and our ability to be happy and fulfilled are heavily influenced by early childhood circumstances and conditions. Although we now accept this insight as common knowledge, fifty years ago it was not common.

Before this new insight, how we parented was not a priority. Our success in life was attributed mainly to genes, family status, hard work, opportunity, character, religious affiliation, or luck. In Eastern cultures, which commonly believe in past and future lives, past karma was also seen to be the major contributing factor. If you were good in a past life, then good things will happen for you in this life. Certainly, parents have always loved their children, but how they demonstrated that love with parenting skills was not recognized to be that important.

Now, after fifty years of counseling psychology, we have discovered the way parents demonstrate their love makes an enormous difference to their children. With this increased knowledge of the importance of childhood, parents today feel much greater pressure and responsibility to find the best way to parent their children. Often this pressure to be perfect parents leads them in the wrong direction.

Parents commonly make the mistake of focusing too much on providing more. And what they are providing more of is often counterproductive: more money, more toys, more things, more entertainment, more education, more after-school activities, more training, more help, more praise, more time, more responsibility, more freedom, more discipline, more supervision, more punishment, more permission, more communication, etc. More of these things are not necessarily what children today need most. As in other areas of life, more is not always better. Instead of more, what children need is "different." As parents, we don't have to give more, instead we need an approach different from our parents'.

REINVENTING PARENTING

Today we are faced with the challenge of reinventing parenting. Instead of assuming responsibility to mold our children into responsible and successful adults, it is becoming increasingly apparent that our role as parents is only to nurture what is already there. Within every child are the seeds of greatness. Our role is to provide a safe and nurturing environment to give that child a chance to develop and express his or her potential.

Traditional parenting skills and approaches that were appropriate in the past will not work for children today. Children today are different. They are more in touch with their feelings and thus more self-aware. With this shift in awareness, their needs have changed as well. Every generation moves ahead to solve the problems of the past, but new challenges emerge in making that step.

In any field of endeavor, we must make adjustments to

continue being successful. The needs of our children today are different from previous generations'. As parents, we are now facing a change that has been in the making for the last 2000 years. It is the shift from fear-based to love-based parenting.

Positive parenting is a shift from fear-based to love-based parenting.

Positive parenting focuses on new approaches and strategies to motivate children with love and not through the fear of punishment, humiliation, or the loss of love. Though this sounds reasonable when compared to traditional approaches to parenting, it is an extremely radical notion. Love-based parenting is in conflict with our deepest instinctive reactions when we feel that we are out of control or when we feel afraid of losing control.

This love-based approach focuses on motivating children to cooperate without using the fear of punishment. Every parent knows the automatic reaction of, "If you don't stop, I will . . ." And then comes the threat. Or the old fashioned phrase, "If you don't listen to me, I'll tell your father when he gets home." Managing our children with fear, no matter how much we don't want to do it, is an automatic reaction. In many schools today, teachers attempt to motivate their children to do better by means of fear of college entrance exams. All this fear just makes our children more anxious or depressed. Some children are already preparing for college in first grade.

Giving up spanking, threatening, and punishing may sound like a loving thing to do, but when your child is throwing a tantrum in the checkout line, and you just don't know what else to do, threatening or spanking seems to be the only solution. When your child refuses to get dressed in the morning for school or resists brushing his teeth at night, automatically you resort to threats and punishment. Even if you don't want to use threats and punishment, when nothing else works it is all you have. And it is all you have, because we haven't yet learned the skills of positive parenting.

When your child is throwing a tantrum in the checkout line, you just don't know what else to do; threatening or spanking seems to be the only solution.

It becomes possible to change our parenting approach and to do it differently from the way we were raised only when we find a new way that works. You can successfully give up outdated fear-based parenting skills when you have learned the new and necessary skills to awaken and draw from your children the part of them that wants to cooperate and is already motivated to behave in harmony with your will and wish.

A SHORT HISTORY OF PARENTING

Thousands of years ago, children were treated worse than we would treat animals today. If children disobeyed a parent, they were severely beaten or punished, and sometimes even killed. Burial sites in Rome from two thousand years ago have revealed the bodies of hundreds of thousands of young boys who were beaten and killed by their fathers for being disobedient. Over the years, we have moved away from such extreme abusive and violent treatment.

Today most parents spank or hit their children only as a last resort, when nothing else seems to work or when the parent goes out of control. Still the legacy of the past holds on. Even in a relatively peaceful home, children can be heard saying, "If you do that you'll be killed" or "They'll kill you for that." Although, if questioned, these children don't literally mean "killed," but it is a clear indication of the influence of fear to create orderly or good behavior. Although we have come a long way in the last two thousand years, the use of fear remains entrenched.

Some parents still think their children need to be spanked. I remember one dramatic example. Ten years ago, I had a conversation with a taxi driver from Yugoslavia. He mentioned that the problem in America is that parents are too soft. They don't beat their children. I asked him if he was beaten. He was proud to say that is why he turned out so well and so had his children. Neither he nor his children had ever spent the night in jail. He went on to say that not a day passed when he was growing up that he was not beaten. As an adult, he was grateful for the beatings he had received. He assured me that this was a common practice in his country and it had saved him from becoming a criminal.

This is an amazing psychological reaction. Quite often, children who are severely beaten or abused will bond even more with the abuser. Over time, they begin to justify the abuse and feel they deserved it. Instead of recognizing what they experienced as abuse, they defend their parents' behavior. When they have children, naturally they feel their children deserve the same abuse. This is why it can be so difficult for some parents to adjust to positive parenting. They hold on to fear-based parenting, because they were punished and feel that their children deserve it as well. They believe their rearing helped them to be better citizens and so it will help their children. It is common to hear an abused child say, "I was so bad that they had to beat me."

Children who are severely beaten or abused
will bond with the abuser and defend the
abusive behavior.

Certainly, many more parents who were beaten now recognize this to be an outdated practice, but really don't know what else to do. Though they don't like spanking or punishment, they don't have an alternative. Other parents have given up spanking, and as a result lose control of their children, or their children develop self-esteem issues. If we are to give up spanking and punishing, we must replace them with something that works effectively to manage children and create cooperation.

VIOLENCE IN, VIOLENCE OUT

When children are receptive, feeling, and open, as children are today, once violence goes in, it comes right out. There is no doubt that when children are managed by using the threat of violence, punishment, or guilt, they will resort to violence, punishment, or guilt when they feel out of control as a way to regain control. All the rampant dysfunctional behavior and domestic violence in our society today is the result of not knowing another way to deal with the strong feelings that people feel.

When feelings were not so awakened, violence and punishment worked. But today the world is different. Parents are more conscious and more feeling and so are their kids. Without a new way of managing and controlling children, they will become increasingly violent and continue to behave in dysfunctional ways. Either they will act out in rebellious and aggressive ways, or they will turn that violence inward and abuse themselves with low self-esteem. Either they hate others or they hate themselves, and often they feel both.

**Children exposed to violence either hate
others or hate themselves.**

I can only laugh when some experts say there are no scientific studies to prove that spanking makes children violent. That is what they said when I began teaching *Men Are from Mars, Women Are from Venus* more than fifteen years ago. They asked, "Where are the studies to prove that men and women are different?" It was just common sense.

Scientific studies are very useful to expand our awareness and insight, but when we become so dependent on scientific studies and ignore common sense we have gone too far. Scientific inquiry then becomes as dangerous as the superstition it helped society to escape. Fortunately, not all scientists and researchers are so narrow-minded that they easily dismiss common sense.

**When we are dependent on scientific
studies and ignore common sense,
we have gone too far.**

Although "violence in, violence out" is common sense, it has also been shown through studies that exposure to violence makes children more violent. After the riots in Los Angeles in 1989, different groups of children were shown videotapes of the violence for three minutes. Afterward, they played in another room where there were violent toys and nonviolent toys.

When told that the violence on TV was just actors pretending to be violent, the children didn't play with the violent toys but played with more neutral or nurturing toys. When told that the violence on TV was real, almost all the children played with violent toys. Aggression dramatically increased. Watching real violence on TV clearly triggered increased aggression in these children.

It is not until age fourteen that children or preteens have the cognitive development fully to understand a hypothetical situation. Even if a child or preteen is reminded that the players on TV are only pretending, he or she cannot remember for long. And if they reminded, after five or ten minutes, they will still emotionally respond as if it was real. Without the required cognitive development, what a child feels is real becomes real. When children witness violence or mean behavior on TV, they lose, to some degree, the opportunity to develop a healthy sense of innocence, serenity, and sensitivity.

**When children are not over stimulated by
violence or meanness on TV, they are clearly
more secure, relaxed, and peaceful.**

If a parent decides a movie is okay for their preteen, but still has some doubts, then it is better to have their preteen watch it when it comes out on video. Video, in the home with the lights on, has much less of an impact than a dark movie theater with a bigger-than-life screen. A theater increases an adult's ability to suspend his or her disbelief and to experience the emotional ups and downs of the movie. Movie theaters are designed for adults to forget what is real, so they can temporarily feel as if the movie is real. For children, we want them to remember that that what they are watching is not real.

Too many movies or too much TV, even without violence and mean acts, can be over stimulating to children. One of the most common reasons children act out or can't control themselves. Children learn primarily by imitation. What they see is what they do. Too much sensory input overwhelms their nervous system, and they become irritable, demanding, moody, hyper, whinny, too sensitive, and uncooperative. Too much stimulation is not a healthy influence.

Yet, many of the very people that complain loudest about violence on TV will turn around and threaten their children with violence and punishment. Yet they are right. Violence on TV and in the movies does influence our children and teenagers, but this conclusion about TV is misleading, because the influence of the parent and their philosophy and practice of parenting is much greater.

Parents have a much greater influence on their children than does TV.

When children are raised to believe they are bad and they deserve punishment, violence on TV and in the movies has a much greater negative impact. When children are raised without spanking, punishment, or guilt, they are still influenced by violent programming, but at least they are less attracted to it. Parents should be diligent in protecting their children from the influence of too much sex and violence in the movies and on TV.

The power to raise healthy children is within the parents' reach. We cannot fully blame the problem of increasing youth violence on Hollywood. Hollywood only provides what we want to see. As long as children are raised with fear and guilt, they will continue to want the violence Hollywood offers.

WHY CHILDREN BECOME UNRULY AND DISRUPTIVE

There are clear reasons why children in schools today are more unruly, disrespectful, aggressive, and violent. It is not a big mystery. When children are overstimulated by aggression or the threat of punishment at home, it creates hyperactivity in boys—or what is now diagnosed as Attention Deficit Disorder. In girls, aggressive tendencies are acted out against themselves with feelings of low self-esteem and eating disorders.

Go into any prison, and you will find that all violent offenders, without exception, have been severely punished or beaten as children. The abuse they have suffered is heartbreaking just as the abuse they inflicted on their victims is heartbreaking. Yet even outside the prisons and in the counseling office, many millions of people suffer from depression, anxiety, apathy, and other emotional disorders as a result of fear-based parenting.

On the other hand, there are many children today who are disruptive and impaired from the affects of "soft" parenting. Traditional parents are correct in being skeptical about modern soft-parenting approaches. Although the intent to be love-based is present, the skills to make it effective are not being practiced. The freedom and power given by the five messages must be balanced by equally powerful skills to maintain control over children and create cooperation. If you want to drive a fast car, you must make sure you have great brakes. You can't give children more freedom unless you have the skills to restrain them so that they behave in an orderly manner.

You can't give children more freedom unless
you have the skills to restrain them so that
they behave in an orderly manner.

Many parents who were mistreated as children resolved never to hit, spank, degrade, or punish their children. They knew what didn't work and, to be better parents, stopped doing it. The problem is they didn't know how to replace the old fear-based practices with love-based skills. Refusing to discipline their children in many cases spoiled their children. This kind of soft parenting is just as ineffective as traditional fear-based approaches.

Giving up past fear-based techniques only works when you replace them with something else that is more effective. Although children today have new needs, they still need a parent who is in control. Otherwise, no matter how much you love your child, the child goes out of control.

Positive parenting uses the practice of making the child take a time out in a variety of ways which are age appropriate to replace the need to spank or punish. Even then, time outs are used as a last measure. Long before resorting to a time out, there are many other skills to be applied so that a time out works. Otherwise, it just becomes another fear-based punishment and loses its effectiveness.

Positive parenting uses the practice of time outs to replace the need to spank or punish.

In light of an alternative way of parenting our children without fear or guilt, we really need to stop and consider why anyone deserves to be beaten or feel pain because they have made a mistake. No one ever deserves punishment. Everyone deserves to be loved and supported. Even in the past, no one ever deserved punishment, but it was the only way to regain and maintain control. Punishment and spanking helped parents keep the upper hand and control their children. Today, punishment and spanking have the opposite effect.

In the past, punishment maintained control, but today it has the opposite effect.

In the past, children did not have the capacity within themselves to know what was right or wrong. The fear of punishment was necessary to deter them from misbehaving. The more resistant children were, the more punishment they received. Punishment was needed to break their will. It is precisely this kind of strategy that would allow people to tolerate and even support the abuses of tyrants and dictators throughout history. Weak-willed people will allow abuse. Fortunately, times have changed and Western society will not tolerate and support abusive tyrants. Just as society has changed, so have our children. Our children will not be broken, but will continue to rebel in response to spanking and punishment.

If you are still against giving up spanking and punishing, ask yourself this question: If there was another way to have the same or even better effect that didn't involve fear, punishment, or guilt, would you consider it? Of course you would. We cling to fear, punishment, and guilt only because we don't know another way. As you read on to learn these new non–fear-based techniques, they will not only make sense, but will also work. That is the whole point. We are not exploring the philosophical pros and cons of parenting approaches. We are talking about an alternative approach that will start working right away.

Thousands of people in my seminars and workshops on parenting have already started to use this approach with success. It not only works, but it feels right in your heart. Let your heart and common sense give you the confidence and courage to move ahead in giving up outdated parenting tactics and begin using these new skills of positive parenting.

A GLOBAL SHIFT IN CONSCIOUSNESS

During the twentieth century, Western psychology developed in response to the new needs of the collective consciousness. Prior to the last hundred years, an introspective exploration of our inner feelings, desires, and needs was not that important. People were more concerned with their survival and security and not worried about how they felt. Most people were not even aware of their feelings. To a great degree, most people were not even aware of their psychological and emotional needs.

Just as the world has changed, our children have as well. Many times, my children are more articulate and aware of their inner feelings than I am of mine. We have all been born at a time of tremendous change in global consciousness. As the collective consciousness of society has shifted, our inner world has become more important. The attributes of love, compassion, cooperation, and forgiveness are no longer lofty concepts for philosophers and spiritual leaders, they are daily experiences. The behaviors and practices of people in power that were once acceptable are now seen to be abusive.

The attributes of love, compassion,
cooperation, and forgiveness are no longer
lofty concepts for philosophers and spiritual
leaders, they are daily experiences.

History is filled with atrocities of human conscience. Throughout the Dark Ages, different religious and political institutions were responsible for brutally murdering and torturing millions of innocent men, women, and children simply because they had different beliefs about God or chose natural herbs to heal their bodies. These atrocities have continued even into the twentieth century. Yet today, most people oppose them. Since human consciousness has evolved, justifying these kind of atrocities has become almost unthinkable.

You don't really need to be taught anymore that killing, stealing, raping, and pillaging is wrong. Your conscience tells you these are not right. No one really needs to convince you. In a similar way, it is unlikely that you would allow a political leader to start a war, dominate a country, and steal all of its precious cultural and art objects. Yet today we have museums around the world that are filled with stolen objects or the "spoils of war." This kind of psychopathic egocentric behavior was acceptable just fifty years ago.

As the collective consciousness of society changes, conscience evolves, as well as intelligence. When people are not capable of knowing what is right or wrong, they need lots of rules which then must be enforced with punishment. If one is capable of developing a conscience, then the need for punishment is outdated. Rather than focus on teaching children what is right and wrong, positive parenting is more focused on awakening and developing children's innate ability to know within themselves what is right and what is wrong.

Rather than teach what is right or wrong,
teach how to know within yourself what is
right or wrong.

Having a conscience is the ability to know within ourselves what is right or wrong. It is like having an inner compass that always points us in the right direction. We don't always have all the answers, but our inner compass will always point ourselves in the right direction. In the past, some people have described conscience as listening to a quiet inner voice. That was the only way they could describe something that most everyone now experiences. We now just say, "I had a feeling."

Feelings are the doorway through which our inner soul or spirit communicates to us. When people are "stuck in their heads," all they can do is follow the rules and punish those who don't. People with open hearts are able just to know what is right for them. This same inner knowing, when applied to interpreting the world, is called intuition. When applied to problem solving, it is called creativity. When applied to relationships, it is the capacity to love (or recognize a person's goodness) without conditions and forgive. Developing the mind is certainly important, but developing a conscience is the most precious gift parents can give to their children.

All parents want their children to know what is right and then on that basis, to act wisely. Until this global shift in consciousness occurred, that was not possible. Parents employed punishment and other fear- and guilt-based strategies to deter children from being bad and motivate them to be good. Today children are born with a new potential to develop this inner knowing. Their sensitivity gives them this ability, but it can cause them to self-destruct when outdated fear-based strategies are employed. Whatever treatment goes in either comes right back out or becomes self-directed.

Children today tend to self-destruct in
response to fear-based parenting.

Every child born today has the innate ability to know what is right and wrong. They have the potential to develop a conscience, but that ability must be nurtured if it is to come out.

Positive-parenting practices awaken that inner potential in our children. The result of being connected to an inner conscience is that our children are well behaved but not mindlessly obedient. They respect others, not out of fear, but because it feels good. They are willing to and capable of negotiating. They can think for themselves. They are willing to challenge authority figures. They are creative, cooperative, competent, compassionate, confident, and loving. By learning and applying positive-parenting skills, not only does parenting get easier and easier for us, but the rewards for our children are so much greater. There is no greater reward in life than seeing your children succeed in making their dreams come true and feeling good about themselves.

3

New Skills to Create Cooperation

The sooner you experience the power of positive parenting, the easier it is to give up fear-based parenting skills. Just give yourself one week to practice the ideas in this chapter, and you will never want to go back. Remember that for positive parenting to work you can't revert back to punishing or threatening your children. You will find that your children will magically begin to respond. This is true for children of all ages. Even your teenagers will respond. The earlier you start, the more quickly your children will respond. When children or teenagers are used to being controlled by fear, it can take a little longer, but it still works. It is never too late to apply these positive-parenting skills. In many cases, they are skills that will help you communicate better with your spouse as well.

ASK, BUT DON'T ORDER OR DEMAND

To create cooperation is to instill in children a willingness to listen and to respond to your requests. The first step is to learn how to direct your children most effectively. Consistent ordering does not work. Think about your own experience at work. Would you like someone always telling you what to do? A child's day is filled with hundreds of orders. It is no wonder that mothers complain their children don't listen. Wouldn't you just tune out if someone nagged you all the time?

A child's life is filled with orders, for example: Put this away, don't leave that there, don't talk to your brother that way, stop hitting your sister, tie your shoes, button your shirt, go brush your teeth, turn off the TV, come to dinner, tuck in your shirt, eat your vegetables, use your fork, don't play with your food, stop talking, clean up your room, clean up this mess, sshh, get ready for bed, go to bed now, get your sister, walk—don't run, don't throw things in the house, watch out, don't drop that, stop yelling—and on and on, again and again. Just as parents become frustrated nagging a child over and over, the child just tunes the parents out. Repetitive orders weaken the lines of communication.

The positive-parenting alternative skill to ordering, demanding, and nagging is asking or requesting. Wouldn't you rather be asked by your boss (or spouse) rather than be told? Not only do you respond better, but your children will as well. It is a very simple shift but it takes lots of practice. For example, instead of saying, "Go brush your teeth," say, "Would you go brush your teeth?" Instead of saying, "Don't hit your brother," say, "Would you please stop hitting him now?"

USE "WOULD YOU" AND NOT "COULD YOU"

Make sure that, when phrasing your request you use the words "will" or "would" instead of "can" or "could." "Will you" works wonders, while "could you" or "can you" creates resistance and confusion. When you say, "Would you clean up this mess?" you are making a request. When you say, "Could you clean up this mess?" you are posing a question about competence. You are asking, "Do you have the ability to clean up this mess?" To motivate cooperation, you have to be very direct and very clear about what you want. You must first present your request in a way that evokes cooperation.

It is fine to say, "Could you clean up this mess?" if you are really asking about their competence. If you are asking a child to do something, be direct. Most of the time parents will say "could you" as a way of ordering their children with a little guilt tossed in. Most often parents do so because that is how their parents behaved and it is automatic. Although it may seem like a little thing, how you ask makes a huge difference in children's willingness to cooperate.

"Could you clean up this mess?" is not a
request; it is an order with a lot of confusing
indirect messages thrown in.

Regardless of intent, when a parent speaks in a bothered, frustrated, disappointed, or upset tone and uses "could you" or "can you," a child may receive indirect messages. If the parent says, "Could you clean up this mess?" the child may hear any of the following messages:

"You should clean up this mess."

"You should have already cleaned up this mess."

"I shouldn't have to ask you."

"I have asked you before to clean up your messes."

"You are not doing the things I have asked you to do."

"You are not acting your age."

"You are a real pain to me."

"Something is wrong with you."

"I am in a big hurry and I can't do everything."

Although none of these messages may be directly intended, it is what children hear. All this indirectness and guilt sabotages the possible results of positive parenting. After you practice these techniques, you will find that directness without guilt or fear is much more effective.

To understand this more clearly, let's imagine you could map the activity in children's brains. When you ask a "could you" question, there would probably be activity in his left brain wondering what exactly you mean. If you use "will" or "would," there would be activity in the right brain and the motivation center would be activated.

Using "will" or "would" bypasses much
of children's resistance and invites him
to participate.

Take a moment to pretend that you are a child hearing either of these two different questions: "Could you go to bed and stop talking?" or "Would you go to bed and stop talking?" At first, it feels like the "could you" phrase is more polite. "Would you go to bed and stop talking?" seems more authoritarian and may be too controlling. Then, as you continue to feel the difference, "could" sounds nice but there is also a hidden order saying, "I am asking you nicely but you'd better do it or else." Then, as you continue to consider "Would you go to bed and stop talking?" it seems to be inviting you to cooperate. If you want to object, you are free to. Clearly, this is the message we want to give our children. When we just order our children, we actually prevent them from learning to be cooperative.

These little word changes make a world of difference, particularly with little boys. "Would" and "will" not only work better with little boys, but also with grown men as well. Women tend to resist asking, and, when they do, they often do it in indirect ways. Not only do men need this directness, but children need it even more.

To use "could" and "can" sends confusing messages and gradually numbs children's willingness to cooperate. You are the parent. You would not be asking her if you didn't already believe that she could do the very thing you are asking. When you ask, "Could you turn off the TV?" you are not really asking if they have the ability to turn off the TV. You want them to turn off the TV, and you're giving an unspoken message that they have no good reason if they don't turn it off.

Although it sounds polite to use "could you" and "can you" to create cooperation, they are ineffective. To repeatedly use "could" and "can" sends confusing messages and gradually numbs children's natural willingness to cooperate.

I began using this technique when my daughter Lauren was a baby, at first because I wanted to prepare all of my three daughters for being in successful relationships later in life. As I stated in *Men Are from Mars, Women Are from Venus,* one of the most important skills women need to learn in their relationships with men is how to ask for support in a way that motivates a man rather than repels him. Women are not taught how to ask for what they want in childhood.

I knew that the best way to teach my children was to model the behavior, so, to teach them, I started asking them to do things with "would you" or "will you." I was pleased that they picked it up so easily. Other parents were quite amazed when in preschool Lauren would say, 'Would you please help me?" or "Would you not talk to me that way?" or "I have had a hard day, would you please read me a story?"

Although my intent was to teach them how most effectively to ask for what they wanted, which they did learn, the side effect I later discovered, was that using "would you" or "will you" made them much more cooperative. Now, when parents begin creating greater cooperation by using "would" and "will" in a clear and direct manner, they are also preparing their children to master the art of asking for what they want and getting it.

GIVE UP RHETORICAL QUESTIONS

Even worse than using "could you" and "can you" is using rhetorical questions. Rhetorical questions are fine when you are trying to make a point in a persuasive speech, but they are counterproductive when asking for cooperation. For every rhetorical question, there is always an implied message. In parenting, the implied message is usually a negative guilt message that a loving parent wouldn't want to say directly. Instead, it is implied in a rhetorical message. Many mothers don't even realize they are giving a negative message, but with a little soul-searching, it is easy to recognize.

Women particularly will use rhetorical questions to motivate children to be obedient. When a mother wants her child to clean up his room, instead of saying, "Would you please clean up your room?" she throws in a little shame and guilt by using a rhetorical statement first like, "Why is this room still a mess?" Let's explore a few examples.

Rhetorical Question	Possible Implied Messages
Why is this room still a mess?	You should have cleaned this room. You are bad. You are a slob. You don't listen to me the way you should, etc.
When are you going to grow up?	You are behaving in an immature manner. I am embarrassed by your behavior. You are a big baby. You should be behaving differently.
Why are you hitting your brother?	You are bad for hitting your brother. You are really stupid. You have no good reason to be hitting your brother and yet you do.
Are you okay?	Something is wrong with you. You are behaving in a strange manner. Unless you have a good reason, you have no excuse for your behavior . . . it is bad.
How could you forget to do that?	You are either really stupid or very bad and insensitive. You are a pain in my life. I cannot depend on you for anything.
Why are you still talking in here?	You should be sleeping. You are really bad kids. I have told you again and again, and you still don't listen to me.

By giving up rhetorical questions before making a request, parents increase their chance of creating cooperation; otherwise children just stops listening. Avoiding rhetorical questions not only helps create cooperation, but it also prevents your children from being exposed to poor communication skills. Rhetorical questions not only don't work for the child, but they prevent parents from clearly taking responsibility for the negative messages they are sending out. Without clearly recognizing our negative messages, it is hard to understand why our children are not willing to cooperate with us.

BE DIRECT

One of the most important skills for mothers to learn is to be direct, particularly with little boys. Women will often state what they are displeased about, but do not follow it with a request. This is like fishing in a desert. They have little chance of getting the response they want. Here are some examples of not being direct:

Negative message	Implied order
You kids are making too much noise.	Be quiet.
Your room is a mess again.	Clean up your room.
I don't like the way you are treating your sister.	Be nice, don't treat her that way.
You shouldn't hit your brother.	Don't hit your brother.
You are interrupting me again.	Don't interrupt me.
You can't talk to me like that.	Don't talk to me like that.
Your shoes are untied.	Tie your shoes.
You were late last time.	Be on time.

In each of these examples, the parent is trying to motivate the child to do something by focusing on the problem but is not asking him to do anything. The implied request is often not even realized by the child, who does nothing but stare into space. To get a direct response, the request needs to be direct without focusing on the negative expression. Focusing on what a child did wrong or why a child should feel bad does not help create cooperation. Let's explore how negative messages could be rephrased as effective requests for action.

Negative message	Positive request
You kids are making too much noise.	Would you please be quiet?
Your room is a mess again.	Would you clean up your room?
I don't like the way you are treating your sister.	Please be nice, don't treat her that way.
You shouldn't hit your brother.	Please don't hit your brother.
You are interrupting me again.	Would you please not interrupt me?
You can't talk to me like that.	Please don't talk to me like that.
Your shoes are untied.	Would you tie your shoes?
You were late last time.	Please be on time.

GIVE UP EXPLANATIONS

Besides making sure to ask instead of ordering or demanding, to motivate children to action, don't give them a reason. Many well-meaning experts suggest focusing on giving children a good reason to do the action. This does not work. As a parent, when you explain your position to justify your request, you give up your power. You confuse the child. So many well-meaning parents try to convince their children to follow instructions instead of reminding them it is okay to resist, but mom and dad are the bosses.

You don't need to say, "It's time to go to bed; you have a big day tomorrow. Would you go brush your teeth?" Just say, "Would you go brush your teeth?" Leave out the explanation. When children resist their parents, they are mostly resisting the reasons. When you leave out the reasons, they have less to resist.

Most men experience this when responding to a woman's request. Often women will give a big explanation why he should do something, when he would much rather that she be brief. The more she talks about the reasons he should do something, the more he will feel resistant. Similarly, the briefer you make the request, the more willing your child will be to cooperate.

If you want a young child to understand why it's good to go to bed, tell the child later, once you are pleased with her for cooperating. After she is in bed, you could say something like this, "I am so pleased with you. You brushed your teeth so nicely. And now you can get lots of sleep to prepare for tomorrow. It's a big day and a good night's sleep will make you feel good tomorrow." When children have done something well, they are much more receptive to little talks.

Most parents give talks to motivate children when the children are resistant, or when they have done something bad or wrong. This kind of timing just reinforces feelings of inadequacy and guilt and eventually disconnects a child from their natural willingness to cooperate. It may appear to work when children are very young, but at puberty, to the extent the child submitted to your will by being a good and obedient child, he or she will need to rebel. To encourage cooperation, giving up explanations will make a big difference.

These are few examples of common mistakes parents make and alternative ways of asking:

Explanation	A Better Way of Asking
You have watched too much TV today; it's time to turn off the TV. I want you to do something else with your time.	Would you turn off the TV and do something else with your time?
Every time we are ready to go to school, you forget where your shoes are. I want you to always put them in one place so that you can remember.	Would you put your shoes in one place so that you can always remember them?
I have been picking up after you all week long. I want you to pick this stuff up right away.	Would you please pick this stuff up right away?
I am really tired today. I can't deal with cleaning up. I want to wash your dishes tonight.	Would you wash your dishes tonight? That would make me very happy.

GIVE UP GIVING LECTURES

Even worse than explanations before a request are lectures on what is right or wrong and good or bad. It is actually counterproductive to say, "It's not nice to hit your brother. Hitting isn't okay. Would you please stop hitting him now?" Besides sounding contrived and unnatural, it just doesn't work. Certainly, stating a rule or policy is okay, but not to motivate a child. When lectures on good or bad are given to motivate behavior, children disconnect from their willingness to cooperate and instead try to figure out right from wrong, good from bad. Children younger than nine years old are not prepared for such heady stuff, and after age nine, they will just stop listening.

The only time to give children or teenagers a lecture, no

matter how old they are, is when they ask for one. Many parents complain that their children don't talk with them. The major reason is that parents offer way too much advice and too many lectures. Children will particularly turn off to lectures when a parent uses them either to motivate them to do something or to tell them why they are wrong. In both cases, lectures are not only worthless but also counterproductive. Here is a sample lecture:

"Your brother did not mean to hurt you. He was just playing and accidentally bumped you. The best way to get along is to use your words and not hit. Hitting will just make the problem worse. If a bigger kid at school were to hit you, it wouldn't feel very good. In a similar way, it doesn't feel good to him when you hit your brother. Better than hitting is using your words. Instead of hitting him, you could have said, 'I don't like being hit, please stop.' If he continues, then repeat your words. Remember you don't have to hit. There is always another way. Sometimes you can just walk away, like you are bored with this behavior. On the other hand, if you want to fight, then I would be happy to oversee a wrestling match or we could put on the boxing gloves. It is good to learn how to protect yourself if there is nothing else you can do to protect yourself, but it is not good to fight with your brother. You both know how to use your words and you can always ask me for help. . . . So don't hit your brother."

Unless a child is really asking for the information, although it may be good and useful, it will just create more resistance.

DON'T USE FEELINGS TO MANIPULATE

Feelings are to be shared among equals. To help train children in the important skill of identifying and sharing feelings, parents make the mistake of using "I feel" statements. Many books are filled with suggestions about always communicating your feelings to children. Although this is well-meaning advice, it is often counterproductive when it comes to creating cooperation.

Parents are taught to use this simple formula to create cooperation.

When a b c, I feel 1 2 3 because I want x y z.

For example, "When you climb the tree, I feel afraid you will fall. I want you to come down."

Or "When you hit your brother, I feel angry because I want you to not hit each other and to get along."

This formula and other similar formulas are effective to assist children in communicating their feelings to each other or to help adults communicate to each other. It does not work to use feelings to cross the generational line. When parents, who are the bosses, share their negative feelings with children to motivate behavior, it makes children feel overly responsible for the parent. The result is that they feel guilty for upsetting the parent and adjust their behavior, or they feel manipulated and resist cooperating. Negative feelings should not be shared with children. It is not appropriate for the "boss" to get on equal footing with the child. As soon as you express your negative feelings, you lose much of your control and the power to create cooperation.

Parents who share their feelings with their children wonder, "Why do my children resist my authority so much?" Eventually, as these children reach puberty, they stop communicating with their parents altogether. Many grown men have a difficult time listening to their wives' feelings, because they felt so manipulated by their mothers' feelings as children. A dramatic example, to make this point clear, is the mother or father who says, "When you do this, I feel so disappointed. (I work so hard so that you have a good life and you don't even try.) I want you to do as I say." This child has only two options: either to feel bad or to stop caring. Neither option is healthy.

When parents are upset and need to talk about their feelings, they should find another adult to find comfort and support. It is not appropriate to look toward your children for emotional support. Certainly, it is great to share your positive feelings with your children, but your negative feelings will just be rejected as a form of manipulation.

Some parents assume if they say "I am really angry" that this intimidation will motivate their child. It certainly will deter a child, but it is fear-based and will eventually numb a child's natural willingness to yield to your wishes. Using feelings to manipulate will tend to make some children obedient but not cooperative. Many children, particularly boys, will just turn you off. They will stop listening and even stop looking you in the eye.

Many parents use "I feel" statements as a way to educate their children into a greater awareness of feelings. This is better accomplished at times when you are not trying to motivate your children to do something. It is best to do in response to their asking how you feel, or if you have ever felt the way they do.

THE MAGIC WORD TO CREATE COOPERATION

Besides being brief, positive, direct, and using "would you" when making a request, one other skill remains. It is the most important. It is remembering to use the most powerful word for creating cooperation. That word is "let's."

To a great extent, until age nine children have not formed a sense of self. To order them around is to reinforce the separation between parents and children rather than work with the natural connection children have to their parents.

Whenever it is possible, invite children to participate with you in some activity. Even if you have asked something specific like, "Would you please clean your room?" you could precede this request or follow it with the phrase, "Let's get ready for the party." By including your request in the context of an invitation to join with you, the result is increased cooperation.

A SHORT REVIEW AND PRACTICE

So far we have explored the foundation techniques of creating cooperation:

Ask but do not order.

Make sure your children feel they are cooperating and not just having to be obedient; give them permission to resist (if and when they do). If they don't have the right to resist, question, or negotiate, then your request is really an order or demand.

Make sure you use "would" and "will" and a lot of "pleases."

Give up rhetorical questions, explanations, lectures, and "feeling" statements.

Make sure to be direct and, whenever possible, be positive.

Whenever possible, use an inclusive statement with the word "let's."

Creating a willingness to cooperate really isn't that hard but it does take a lot of practice. It becomes easier by just focusing on making short requests instead of orders. Use the following list to practice

Ordering	*Requesting*
Put this away.	Let's clean up this room. Would you put this away?
Don't leave that there.	Let's now put our things away. Would you please put that away?
Don't talk to your brother that way.	Let's remember to be respectful. Please use a nicer tone when you talk to your brother.
Stop hitting your sister.	Please, right now, stop hitting your sister. Let's all try to get along.
Tie your shoes.	Let's get ready to go. Please tie your shoes.
Button your shirt.	Let's look our best. Would you button your shirt please?
Go brush your teeth.	Let's start getting ready for bed now. Please go brush your teeth.
Turn off the TV.	Let's make sure we're not watching too much TV. Please when this show is over in ten minutes, turn off the TV.
Come to dinner.	Let's come and eat. Please come to dinner.
Stop talking.	Let's be quiet and listen to your mother. Please stop talking.
Eat your vegetables.	Let's remember how important vegetables are. Would you please eat your vegetables?

| Use your fork; don't play with your food. | Let's remember our table manners. Please use your fork and not your hands. |

WHAT TO DO WHEN CHILDREN RESIST

When you first begin this new approach, it will give your children a lot of power. They may laugh at you and say no. Don't worry—this is supposed to happen. They will either be happy to cooperate, or they will be happy to resist. After all, do you always do what you are asked to do? I hope not.

Using the "let's" word is usually fine in most situations until a child gets to the age of nine. At that time, it is a little hokey to say, "Let's clean up this room," unless you are also doing some of the cleaning. Remembering to use the magic word "let's" requires practice, but eventually it will become second nature.

When children resist your initial request, then it's time to move on to step two. These skills of step one are necessary for creating the foundation of cooperation. The skills of step two are needed to motivate your children when they resist your initial request. After much practice, as your children get accustomed to the different positive-parenting skills, these first skills of step one will be more effective. In the beginning, if children are used to being controlled by fear, this first step is needed to lay the foundation for steps two, three, and four. Later on, you will find that most of the time you will only need to ask, and your children or teenagers will cooperate. In the next chapter, we will explore step two and learn new skills for minimizing resistance by understanding our children. When children resist cooperating, the next step is to minimize their resistance.

4

New Skills to Minimize Resistance

Giving your children permission to resist ensures cooperation and not mindless obedience. Although you may feel at times that you would prefer a little mindless obedience to their resistance, there are new skills in positive parenting to minimize resistance. Some resistance from your children is good. To nurture the spirit of cooperation, children need to experience again and again that you are listening to them, just as they are listening to you. Children form a clear and positive sense of self primarily through occasional resistance to your requests.

Children need to experience that you are
listening to them just as they are listening
to you.

Children's resistance, when properly supported by a parent, gradually helps them develop an awareness of their inner world of feelings, wishes, wants, and needs. Ultimately, it ensures that children maintain and develop a strong sense of will. This willpower may be the difference between success and failure later in life. Those with strong wills are able to succeed and those without it will become quitters. Adults with weak wills were not prepared as children to meet and over-come life's inevitable challenges. They settle for mediocrity instead of feeling motivated to make their dreams come true.

FOUR SKILLS TO MINIMIZE RESISTANCE

Instead of demanding obedience, positive-parenting skills use children's resistance to strengthen their will to cooper-ate. Repeated attempts to break a child's will through the threat of punishment or disapproval ultimately undermine a child's natural willingness to cooperate. As long as the will is nurtured and not broken, children's willingness to cooperate will grow and resistance be minimized.

Repeated attempts to break a child's will
undermine a child's natural willingness
to cooperate.

By nurturing our children's need at times of resistance, we can most effectively minimize resistance while keeping their will intact. These are the four ways of nurturing:

1. Listening and understanding

2. Preparation and structure

3. Distraction and direction

4. Ritual and rhythm

To let go of their resistance and feel their inner urge to cooperate, children need understanding, structure, rhythm, and direction. Unless these different needs are being met, children easily disconnect with their inner willingness to cooperate. For example, by means of new listening skills, a parent is able to show that children's feelings, wants, wishes, and needs are being seen, heard, and understood. When this need for understanding is met, children automatically become less resistant and more cooperative.

Although these needs are universal for all children, every child is unique and may have a greater need in one area or another. If a child needs more understanding, it does not mean that he or she doesn't have other needs as well. Each is important for every child, but one or two may be more important for a particular child.

One of your children may respond well to listening and understanding, while another requires preparation and structure. As you become familiar with each of these skills, you will discover how powerful each is. Fulfilling certain needs will create an immediate positive response in your children depending upon their unique temperament.

THE FOUR TEMPERAMENTS

There are four different temperaments in children, which is why they sometimes respond better to one approach rather than another. These four temperaments help to identify your child in a general category and then direct you to employ one of the four skills for minimizing resistance. Although a child may be predominately one temperament, it does not mean that he or she doesn't have a little of each. Some children have equal amounts of each of these four temperaments, while others are more of one and less of another. No temperament is better or worse; they are just different. Since possible combinations are endless, every child is unique and special. Let's explore these four general temperaments.

Sensitive Children Need Listening and Understanding

The first temperament is sensitive. Sensitive children are more vulnerable, dramatic, and feeling. They are acutely aware of how they respond to life in relation to their needs, wishes, and wants. To adjust to life, they have a greater need to identify what they are feeling and then they are more willing to make a change. They respond most to listening and understanding.

Although all children need understanding, these children need it more to release their resistance. Sensitive children learn about themselves by identifying their wants and sharing their feeling responses to life. Complaining is part of their nature. When given the opportunity to share their burdens, they lighten up.

For example, a child might say, "No one even said hello to me. I had a horrible day."

Then the parent says, "That's mean."

Then the child says, "Sarah was really nice to me. She liked the picture I drew."

With a little validation, these children begin to see the positive again. They need more empathy, validation, and recognition of their inner pain and struggles. In general they tend to need more time. Trying to rush sensitive children will create more resistance. They have a different inner clock.

It would be a big mistake for the parent to say, "If Sarah was really nice to you, then maybe it wasn't such a horrible day."

The child would then say, "It was. No one likes me."

When a sensitive child makes the shift to a more positive attitude, just let it be. Don't use this shift as a way to discount the feelings they expressed first.

Sensitive children need empathy and
validation of their pain and struggles.

Without regular messages of empathy, sensitive children begin to dramatize their problems to get the empathy they need. If saying "I have a stomachache" doesn't get a warm attentive response then it becomes, "I have a really bad headache and stomachache, and no one is ever nice to me." Without understanding, every ache is magnified. The lack of empathy will actually create more pain physically and emotionally. When parents ignore a sensitive child, the feelings and problems just get bigger.

The biggest mistake a parent can make is trying to cheer this child up. When they are upset or seem depressed, it does not work to explain why they should not be upset. Focusing on all the positive things may cause them to go in the other direction and focus even more on the negative in an attempt to feel understood and validated. Parents must be careful to listen more and hold back from trying to solve their problems in an attempt to make them feel better.

The biggest mistake a parent can make is
trying to cheer this child up.

Sensitive children need to know they are not alone and that their parents also suffer. This is a very delicate subject. It is not healthy for parents to come to them for emotional support, but parents can share some of their struggles to sensitive children.

For example, after a child complains about how hard, difficult, or painful something is, a parent might say, "I know, why just today I felt really awful, too. I was stuck in a huge traffic jam." Without looking to the child for comfort, this approach satisfies a particular need of the sensitive child.

Sensitive children need to experience that
they are not the only ones who suffer.

When sensitive children resist, they need to hear such empathetic statements as "I understand you are disappointed; you wanted to do this and now I want you to go here." Without the right kind of support, sensitive children cannot let go of their resistance. Without empathy, they tend to feel like victims and can get lost in self-pity. They tend to think more deeply about their suffering, and without being understood, they easily assume blame.

These children need clear messages that their negative feelings are okay. They take a little more time to get over hurts and emotional upsets. Yet, sharing their inner burdens and misery to a sympathetic ear is actually a pleasurable relief. Parents who are less sensitive often mistakenly assume something is wrong with their child and make matters worse.

After hearing the feelings of a sensitive child, give him time and a little space to feel better. When he feels better, don't put a lot of attention on this shift. Have an attitude that is accepting of this temperament as normal and natural. Don't give the message that something was wrong and now he is okay. He was always okay.

These children often resist being pushed into new relationships. Forming relationships and friendships generally takes more time than for other children. They need more help in creating opportunities to meet people and form friendships. When they do establish a friendship, they are very loyal, and when they are betrayed, they are very hurt. Learning to forgive and forget is an important skill they need to learn. When parents listen to their resistance and give understanding, it helps these children to adjust to life's disappointments and increases their ability to forgive.

When these children get what they need, their special gifts can unfold. They are thoughtful, deeply perceptive, creative, good communicators, and original. They are nurturing, compassionate, gentle, and helpful. They derive great fulfillment by serving others and the world.

Active Children Need Preparation and Structure

The second temperament is active. Active children are less concerned with their inner responses to life and more interested in having an influence. They are concerned with doing, action, and results. They are self-motivated and most cooperative when they know what to do or have a plan. They are always ready to move on, lead, or do things their way.

These children need a lot of structure, otherwise they easily go out of your control and resist your authority. They always need to know in advance what the plan is, what the rules are, and who is the boss. They need a game plan. With this kind of preparation these children become very supportive and cooperative. To minimize resistance with these children, you need to think ahead and prepare them with clear limits, rules, and direction.

A parent could say, "This is what we are going to do. First we will play on the swings and then we will go over to the jungle gym. Each of you will get two-minute turns and then we will switch." By preparing this active child with clear structure, he or she will be most cooperative.

Active children always need to know in
advance what the plan is, what the rules are,
and who is the boss.

Active children like to be the center of attention and be
where the action is. They always want to be right. Without
parental structure, they tend to be domineering. This child
needs opportunities to be a successful leader. They respect and
follow a confident and competent leader. A parent must be
careful not to show weakness, indecision, or vulnerability.

For example, don't directly ask them what they think is
best. If they resist by saying what they want after giving
instructions, acknowledge their suggestion and then decide
what is to be done once again.

If you say, "We are first going to the swing and then to
the jungle gym," they might say, "But the jungle gym is more
fun, let's do that first." Then a wise parent could say,
"That's a good idea. Let's do that." These children love to be
right and thrive whenever they are acknowledged.

To minimize resistance from active children it is best to
make them first and/or put them in charge of something
whenever possible. They have a lot of energy and need
parental structure if that energy is to be expressed harmoniously. They are highly motivated to please if given a position of responsibility.

To minimize resistance, make the active child
first or put him or her in charge of something.

Active children need to feel needed and that you trust
them. In this case, the parent could say, "First, we will go to

the jungle gym and everyone gets a chance to get to the top. Billy, I want you to be in charge of making sure everyone gets to the top at least once. You can go first and show everyone the way."

Putting active children in a leadership role with clear guidelines brings out the best in them. Automatically, they become more cooperative. They have lots of energy and feel frustrated when they have to sit still too long. They just have to do something and then will act without thinking and get into trouble. This is why they need structure. When their activities are thought out for them, their abundant energy can flow freely without getting them into trouble.

One way to minimize resistance with active children is simply to wear them out. For example, if you have to wait somewhere, this child will become very frustrated. Either give the child a job to do or create a game to use up the energy. You could have the child simply run a defined distance and back and time the run. Active children love breaking their own records. Give them lots of acknowledgment for their achievements and their resistance will melt.

Active children learn about themselves by doing and making mistakes. They need lots of acknowledgment for their successes and forgiveness for their mistakes. These children have a greater tendency to get into trouble. If they are afraid of punishment or disapproval, they will hide or defend their mistakes, and therefore, they will stop learning and growing from them.

Active children need lots of acknowledgment
for their successes and forgiveness for
their mistakes.

Active children have a difficult time just sitting and listening. They need to move around and learn best by doing and participating with others. When they are resistant to your requests, it is best to begin the activity and invite them to join in. Long conversations are counterproductive and often are regarded as a punishment.

For example, if they resist cleaning up their room, then briefly acknowledge what they are doing and then begin cleaning the room. Say something like, "I can see that you are playing and you don't want to clean up your room. Let's do it together. This is how we do it . . ." Waiting around for them to do something is not very effective.

Active children let go of resistance by joining in with you. Even if they only help a little, thank them for their help and acknowledge how good the room looks. You might say, "We did a good job!" Active children always want to be a part of the winning team. There is no greater motivator than success itself.

**Active children always want to be a part
of the winning team.**

Active children know themselves by what they have done and their results. They like power. When they resist your requests, they often need a firm but calm message that it is okay to resist, but mom and dad are the bosses. Acknowledge what they are doing and then ask again more directly.

For example, you might say, "I can see you are resting in your bed *and* now I want you to start cleaning this room." If the child does not respond, then join in by starting to clean and saying, "Let's start with this part of the room." This approach in sales is called the "assumptive close." You assume your client is with you and you begin exploring the finer details of the purchase.

Active children need a clear and direct message of what you want. "I want" statements minimize resistance by reminding this child that you are the boss. Without the right kind of support, active children tend to go out of control, misbehave, and become abusive to others. Besides structure and supervision, they need clear messages that it is okay to make mistakes and that you know they are always doing their best.

"I want" statements minimize resistance
by reminding the active child that you are
the boss.

When out of control, unless they get everything their way, they will tend to bully others or throw big tantrums. Many parents and adults are afraid to confront this kind of child. Parents put off confrontations because they require so much energy. This just makes the problem worse. Besides clear structure, these children need the structure of regular time outs. They need to feel contained more than other children do. By giving these children regular time outs, they will remember who is boss and get the structure they need. We will explore how to give time outs in greater detail in Chapter 6.

Active children have a greater need to be right and hate being told they are wrong. It is particularly difficult for them to receive corrective feedback in front of others. If feedback is given privately, they resist less and don't become defensive. Instead of correcting them publicly, you can create secret signals to give feedback. They greatly appreciate a parent who will help them to save face.

For example, you could let the child know to be gentler by pulling on your own ear. When they start getting too loud, you could touch your chin to let the child know he or she is using their outside voice while indoors. These children appreciate these signals. It not only helps them be more successful, but it also indirectly acknowledges that it is normal to make mistakes and occasionally go out of control.

Unless given a role of responsibility, they will tend to resist other children or people who move more slowly. They want things fast and have the energy to get them fast. They can best accept the slower pace of others when they are busy doing a job to help or assist in some way. It is even fine to make up activities and give them a sense of importance.

When these children get the structure they need automatically, they become more sensitive, compassionate, and generous. With regular time outs, they gradually learn to be more patient and develop the ability to delay gratification. They become responsible, competent, and make great leaders. They make things happen. Over time, as they feel more successful and confident in themselves, they become more sensitive in understanding others' feelings.

Responsive Children Need Distraction and Direction

The third temperament is responsive. Responsive children are social and outgoing. They develop a sense of self from their responses to the world and their relationships. They are self-motivated to see, hear, taste, and experience everything life has to offer. They have many interests, because they have a greater need for stimulation.

Each new experience brings out a new part of themselves. They come alive in response to new input. Although these children like change, they resist having to focus. They often throw tantrums when being asked to put on a coat or do something in a particular way. They have a greater need for freedom to do their own thing.

Often they don't complete things and just move from one experience to another. It is important for parents to understand this and not worry. This child needs to move around. Chaos is a part of their learning process. Later in life, if they have the freedom to explore, change, and be themselves, they will become more focused and will learn to go deeper into things and complete tasks.

Responsive children naturally move from one activity to another like a butterfly. They need time to explore, experience, and discover life. They are so easily distracted that they need a lot of direction about what to do. When they forget your instructions, they are not trying to annoy or resist you. They have really forgotten. They should never be shamed for this tendency. Gradually they will learn how to stay focused. They are easily distracted by new opportunities. This tendency to be distracted can actually be used to minimize their resistance.

Responsive children know themselves by reacting to life's different experiences.

When resistant to your requests, responsive children simply need to be redirected to another possibility, a new activity, or a different opportunity for experience. Instead of understanding or structure, this child needs to be distracted and then redirected. With distraction, another part of them emerges that is willing to be cooperative. Let's explore a few examples.

When throwing a tantrum, children up to about age

three can be easily distracted and then redirected by pulling out pretty, shining objects, keys, a toothbrush, little shells, a crystal, or anything interesting to see, hear, taste, touch, or play with. My wife, Bonnie, used to always carry a selection of little "things" to distract a child from becoming upset or throwing a tantrum. While this works for all children, it works especially well for this responsive child.

THE GIFT OF SINGING

The activity of singing will distract many children of all ages from what is bothering them and redirect them to feeling loved and supported. Children love when you sing to them and with them as they get older. My wife, Bonnie, created a special little song for each of our children. When they would cry, we could just sing the song and they would feel peaceful again.

As an example of how to do this, I include one of these songs:

> *Lauren Beth, Lauren Beth, how I love my Lauren Beth.*
> *Lauren Beth, Lauren Beth, how I love my Lauren Beth.*
> *Lauren, Lauren, Lauren Beth . . . (and then it repeats)*

When children are distressed, a simple song used over and over will distract them from their troubles and redirect them back to feeling loved and comforted. Singing a song is better than listening to music, because singing actually connects the child more to the parent, although using music in the background to create a more relaxing or happy environment can still be very helpful.

Singing has less gravity or heaviness than the spoken word. It helps to redirect a child from focusing on something that isn't pleasing to something that is. It is ideal for distracting a child and redirecting her to do what you want. You can't stay frustrated and sing at the same time. Singing and listening to song lightens things up and makes life more fun. It stimulates more creative right brain activity. Creativity makes a child more fluid and flexible to situations as well as more cooperative.

When my children were growing up, we used to sing a song while we all washed the dishes together. I called it the five-minute cleanup. I would sing a particular song while we would race to see how much we could get done in five minutes. Afterward, I would acknowledge them for their help and finish up myself. They loved it and still remember it today as a fun and happy experience.

MAKING CHORES FUN

By singing with my children, they were distracted from the drudgery of doing dishes. Also, by limiting their participation to five minutes, they didn't feel burdened. By making sure they didn't have to work too hard at a young age, they didn't resist helping out. As adults, they now are happy to work hard, and they know how to have fun as well.

I can still remember, as a child growing up in a family of seven children, my night to do the dishes. It didn't matter that I only did them once a week. On my night, the feeling I had was, "I always do the dishes. I never get to have fun. Everyone else is having fun and I am missing out."

Children live in the eternal now. When a chore takes too long, it feels like, "Work is all we ever do." By making chores easier and helping your children more, they will learn first how to have fun, and then later in life, as teenagers, schoolwork will be more enjoyable.

Ideally, children should feel taken care of up to seven years old and then between the ages of seven and fourteen, children should focus on having fun by playing, singing, doing crafts, painting, learning a musical instrument, sports, drama, school homework, and doing fewer chores at home. Helping with dishes, cleaning, and taking care of pets is certainly fine, and does not come under the category of "too much work." The best way to determine how much work to give your children is to listen to their resistance and reconsider. Being a parent requires making adjustments every step of the way.

When children learn to be happy, as teenagers they are ready to buckle down and work hard. When children are required to work as children and not have fun, they never learn how to have fun. As teenagers, they often rebel against work or they work hard but have no fun.

Only a hundred years ago, children worked in factories, but gradually society realized that this was abusive. Today we need to realize that it is also a mistake to work our children at home. It is the parents' job to give and the child's job to just receive. Then, around age seven, children have a new need: the need to have fun and play with family and friends. This is the time when children are supposed to develop one of the most important skills in life: the ability to be happy.

It is the parents' job to give and the child's job to just receive.

Most adults really never learn how to have fun and enjoy their lives. This is because they didn't get the necessary support to learn how to have fun. Happiness is a skill, and this skill develops between ages seven and fourteen. Too much academics or too much responsibility and work before puberty will restrict a person's ability to be happy later in life. They either reject work and seek to have fun and be irresponsible as teenagers, or they work hard, but, because they are too serious, they are never happy or satisfied.

Most parents mistakenly believe that they need to teach their children to work hard and be responsible. Children learn to be responsible by having responsible parents. Children learn to work hard by observing their parents work hard. Children learn everything primarily by imitation. They eventually do what they see their parents do. With this insight, parents can have the confidence to follow their hearts and create a happy and fun childhood for their children. Hard work is not needed until puberty.

THE GIFT OF READING

When a child is restless and fussy before going to bed, besides singing little songs, reading stories is an excellent way to prepare the child to relax peacefully and sleep deeply. Although reading to a child before bed is probably the most important gift a parent can give a child, it is particularly important for the responsive child. They hunger to hear stories, myths, and legends. They need the stimulation of far-away places, people, and things.

Children live in a magical world until they are about nine years old. Already society is rushing them to wake up and experience the real world. Parents should not worry. Let your child take time to develop and he or she will easily adapt to the real world when ready. Until they are about seven years old, children do not even have the capacity for logic, and they cannot comprehend an abstract thought until age thirteen.

Let your child take time to develop and he
or she will easily adapt to the real world
when ready.

When children hear on the news that there is a killer on the
loose, they assume they are in danger just as everyone else is.
Using logic to minimize these fears doesn't work. It doesn't
work to say, "Well we have a safe neighborhood and so you
are safe." Magical thinking requires magical solutions. Saying
a prayer for your child's safety will do the trick. If you don't
pray, then wave a magic wand to reassure your child that he or
she is protected. To minimize resistance it is best not to allow
children hear or watch the news up to the age of seven.

By hearing stories, children are easily distracted from
life's burdens. Children use images created by hearing stories
to develop their imagination, creativity, and a stronger sense
of self. Successful people have the sense that they create their
lives, while less successful people feel more victimized or
tossed around by life's challenges and setbacks. Through
increased imagination and creativity, a child is better prepared to solve problems later in life.

Responsive and sensitive children tend to be more at
effect already. By creating their own inner pictures in
response to stories, children develop a stronger sense of
themselves and their ability to create, and will naturally feel
in life more "at cause." Too much watching TV or movies
can weaken this process of creating internal pictures.

USING DISTRACTION TO REDIRECT

All children can benefit from distraction. Up to about eight
years old, children are easily distracted from their resistance

by being told a little story with lots of imagery, color, and shapes. It doesn't matter if the story is not connected to what your child is resisting. Just shift the subject and begin telling a story with descriptive phrases.

For example, when your child is resisting putting on his coat, break out of the power struggle by stopping and talking in story tone. Say something like this, "Oh, look at the beautiful green leaves of that tree. I remember once walking in a beautiful forest and there were giant trees on every side. The sky was such a beautiful blue. There was one big puffy white cloud right above me. I walked all day until I was really tired. It was a long walk but it felt good. Now let's put on this coat."

This is called establishing rapport and then inviting participation. When you tell a story with lots of colors and objects to visualize, the child automatically leaves his or her resistant self and goes into rapport or harmony with you. As a result, the child is much more willing to cooperate.

To minimize resistance, establish rapport and
then invite participation.

At any age, when children are upset and resistant, they respond well to redirection like, "Now let's do this . . ." or "And now we will . . ." Rather than ask the child what she wants to do or even what she would like to do, the parent needs to lead the child. As the child forms her own wishes and wants, she will resist and let you know what she wants. At this point, the parent can say, "Okay, that's a good idea. You do that." Instead of directly asking her what she wants or what she would like to do, make suggestions and let her agree or disagree indirectly through her acceptance or resistance.

Let's explore an example:

MOTHER: Jimmy, let's go play at the park.
JIMMY (eight years old): I don't want to go to the park.
MOTHER: Why not?
JIMMY: I want to play in my room.
MOTHER: Okay, if you want to stay home it is all right
    with me. Take out your colors and make a picture.
JIMMY: I don't want to do an art project. I would rather
    play with my new model airplane.
MOTHER: That's a great idea—you play with your new
    model airplane and I'll come in a little later and see
    how you are doing.

In this way, without asking the child directly, the parent
makes suggestions the child can resist, gradually becoming
able to identify what it is he wants. As a general rule for all
temperaments, it is not a good idea to ask children what
they want, like, need, think, or even how they feel. Instead,
suggest and they will either accept or resist. By resisting,
they form a clear idea of what they want, feel, and think.

It is not a good idea to ask children what
they want, like, need, think, or even how
they feel.

Responsive children tend to be more joyous, light, and
eager. They are literally fed by life's images and changes. For
them, life is an adventure. They tend to be more social and
talk a lot. They make friends easily and tend to like everyone. They are often irresistible, charming, and accommodating. They don't hold grudges.

They don't have deep attachments and are not easily hurt. They have tantrums and resistance, but these usually arise when they are being required to focus or do something they don't want to do. Both chaos and emotional ups and downs are a common part of their life.

They tend to be scattered, forgetful, and unreliable. It is difficult for them to pick up after themselves. They need to be asked again and again. With this insight, parents can find greater peace. Don't expect this child to create order in his or her environment. That is your job. For example, responsive children will not keep their room clean unless they get help. Instead of fighting them, just work with them.

When given the opportunity to have fun and explore many things briefly, they develop their attention span and learn to focus and go deeper. Over time, they will learn to finish tasks. After the age of seven, they need firm encouragement to stay focused. This is most easily done by taking the time to help them.

Without the right kind of support, responsive children, overwhelmed by life's responsibilities, tend to become easily irresponsible or overly scattered and they often reject the responsibilities of being an adult. When they do get the support they need, they become solid, responsible, self-directed, focused, confident, and accomplished.

Receptive Children Need Ritual and Rhythm

The fourth temperament is receptive. Receptive children are more concerned about the flow of life. They want to know what will happen next and need to know what to expect. When they understand the flow, they are most cooperative.

New situations where they don't know what to expect

will trigger resistance. They know themselves by what they expect will happen. When this child expects to be loved, then they feel loved. They need a lot of routine, repetition, and rhythm.

There needs to be a set time to eat, a time to sleep, a time to play, a time to spend special time with mommy or daddy, a time to pick out tomorrow's clothes, etc. They respond well to reassurance and encouragement like, "Now it's time to do this . . ." or "Now we are going to . . ."

They are the most good-natured and thoughtful children. They need more time to do things in an orderly pace and are most resistant to change. They can't make quick decisions and should not be asked what they want, think, or feel. Instead, they need to be told what to do.

Receptive children are the most good-natured and thoughtful children.

As long as it is not a big change, they are most cooperative. A change in time signals a need to change an activity. Using the phrase, "Now it's time to . . ." reassures this child that everything is unfolding and moving as usual. They need things to be predictable around them.

Although they like being told what to do, they will resist being pushed into things or rushed. Like the sensitive child, they need more time to do things or make changes. They need lots of reassurance that everything is preplanned and thought out. It is what they are used to. Repetition gives them comfort. They don't move as much as other children. Often they are just content to stay still and just enjoy being, resting, eating, looking, listening, and sleeping.

They can simply enjoy the passage of time. They are not

automatically self-directed, creative, or innovative. They need to be told, "Now it is time to . . . ," otherwise they may just sit and daydream. They love physical comfort and, rather than risk discomfort, would rather sit and watch.

Unlike the active child, they don't need to lead or even participate. As young children, they often need to just watch and observe. They may watch other children do an activity fifty times and then suddenly just do that behavior. Observing is enough participation for them to develop interest.

Receptive children participate by observing.

A four-year-old child intently watching other children play is not feeling left out. That child is perfectly content to watch. It is as though the child is doing the activity through others. This is not a problem. Eventually, he or she will participate. Around age seven, it is fine to encourage participation, but if a child resists, let him be.

A way to encourage is not by asking, "Would you like to join in?" Instead, use the phrase, "Now it is time for you to join in."

If the child resists, then say, "Okay, I can see you would rather watch. Let me know when you want to join in."

Receptive children are often neglected because they are so quiet, easygoing, and nondemanding. They also occasionally need to struggle and resist. They need to be gently motivated to do things and be challenged, even though they would rather sleep or stay at home.

You have to give these children a task. Unless they get this kind of help, they may not develop any interests. The security of regular routines, rituals, and rhythm supports them in gradually taking the risks required to do something new.

As a rule, if there is a change,
they don't want it.

The responsive child's not wanting to do something is not a good enough reason not to do it. They will never want to do something new. When they resist doing new things, be gentle and never force participation. Remember, by just watching, they are participating on a level comfortable for them. Persist in giving them occasional opportunities to expand their interests, but don't push them to participate. Watching and observing is always good enough.

These children don't like to be interrupted. They want to keep going to the last detail. Repetition gives them security. They will resist when you try to stop them, but their resistance is often silent. They hold back tantrums, because they don't want to cause a problem or be an inconvenience. They have a great fear of disappointing their parents or being rejected.

LOVING RITUALS

Receptive children feel loved by expecting it. Loving rituals need to be created so that these children have a way to experience their worthiness and special connection with each parent. Rituals don't have to take a lot of time; they just need to be recognized as special and then be repeated again and again.

With my daughter, Lauren, we had a special ritual of walking into town through the forest and then resting and having a Madeline cookie at the local bookstore. When she was a young child, I would take her in the stroller and then later, when she was older, we would walk or ride bikes. The whole ritual would take about twenty-five minutes. A ten-minute walk each way and five minutes to eat our cookies and pet the local dogs.

Today, as a teenager, she clearly remembers these early childhood experiences and the loving connection we shared. Many adults have difficulty remembering the love and joy in their childhood and this is a great loss. Being able to remember feeling loved and supported gives us a deep level of security for the rest of our lives.

By simply taking normal activities of parenting and doing them at a certain time, they can become a ritual and thus be most easily remembered. A ritual is created by also talking about it several times. For example, say "Today is Saturday and we can walk into town and get a Madeline cookie." To feel special, children need special activities at special times. Here are some random examples:

- On Saturday morning Daddy is going to make eggs on toast in his special way.

- On Sunday morning we are all going to sleep late and Mommy is going to make her delicious waffles.

- When Daddy is late making a pickup, he always makes it up by taking us to have a smoothie at the grocery store.

- When Daddy is out of town, he always calls to help with homework or say good night.

- Mommy always reads a story before bed.

- Mommy or Daddy always sings a song before bed.

- When a child is sick with a tummy ache, Mommy always prepares the warming pad and puts on a little castor oil.

- When Daddy is happy, he always sings his favorite song.

- Every Thursday at eight, we all get together and watch a very funny family show.

- Every night before bed, the child and parent talk and review their day.

- In spring, the whole family goes to pick flowers before dinner every night.

- Everyone takes the dog on a walk before dinner.

- Every summer we go on a special vacation that we really love, to the same place and the same hotel. (Certainly you could take other trips as well. But by repeating the same vacation, it becomes a special ritual.)

- On Sundays we take a family outing, go for a walk, or have a picnic.

- In summer on Sundays, we go to the beach.

- Every July we go to the county fair.

- Once a month we get to spend the day one-on-one with Mommy or Daddy.

- We say a prayer every night before bed and Mommy or Daddy sings a lullaby.

These fun and loving rituals create special memories and expectations that provide enormous security during childhood and the rest of life. There are other rituals that are less fun or special, but provide a strong sense of security and, most important, rhythm. Everything in nature has a rhythm. Spring follows winter; summer follows spring.

Everything in life has its season. There is a time to be active and a time to rest; a time to eat and a time to play; a time to start and a time to clean up. The tides of the ocean come in and then go out. The sun rises and then it sets. Even in our bodies, we breathe in and then we breathe out. We awaken and then we sleep.

All repetitive behaviors, routines, and rituals provide a sense of rhythm to life. We are comforted by knowing what is coming next. We are familiar with what is to be. All children need rhythm and ritual, but receptive children depend on it the most if they are to come out of their cocoon and express their inner gifts and talents.

PRACTICAL RITUALS

These are more examples of important rituals to provide rhythm to a child's life. Of course, not all of these would be possible or appropriate in every home. They are listed to stimulate ideas.

Get up at the same time every morning before school.

Have your own special spot or chair where you eat.

Go to school at the same time every morning.

Have the same person pick the child up every day after school.

Keep the same schedule to pick up the child.

Go to the park every Tuesday and Thursday.

Wash the car on Saturday.

Come to dinner at the same time; have a special way you call your children to dinner every night—bell or intercom with a simple message, "Dinner is ready. It's time to come to dinner."

Pick out tomorrow's clothes the night before. (This one suggestion helps so much when a child resists getting dressed in the morning.)

Create a little routine before bed to wash face, brush teeth, and put on nightclothes, and start it at the same time every night. (This kind of rhythm is essential. All children need sleep, and a regular routine of getting ready for bed, at a certain time, makes them sleep much better, which in turn makes everything better.)

When receptive children get the rhythm they need, they develop great strength and organizing abilities. They can create order and maintain it. They are peaceful and practical and can overcome great obstacles to achieve their goals. They are very talented at comforting and consoling with loving support. They move slowly, but are very grounded and solid.

GIVING OUR CHILDREN WHAT THEY NEED

Right now, you may be thinking, it is fortunate that parents today are deciding to have fewer children. If this new view of nurturing your children's needs seems like a lot more than you can provide, it is not. It may seem overwhelming because it is new. As you become more familiar with these ideas and begin putting them into practice, parenting will become easier.

These different methods of nurturing do lessen resistance, but they take time and preparation, and sometimes we have neither. The next skills we will explore work even when you don't have a lot of time. In the next chapter, you will learn how listening and expressing your wants can minimize resistance and successfully motivate your children to cooperate.

5

New Skills for Improving Communication

The most important skills for minimizing resistance and creating cooperation are listening and understanding. When children resist cooperating, some part of them is wanting or needing something else. This unmet need, want, or wish must be identified and addressed. Identifying a need or want will often minimize children's resistance. In this case, the act of understanding the source of the child's resistance is enough to make it go away. By learning new skills for improving communication, you can immediately lessen children's resistance and strengthen their willingness to cooperate.

Understanding your child's resistance is enough to make it go away.

The prime directive—or strongest desire, wish, and need—of children is the will to cooperate, please, and follow their parents. Children are born needing to follow their parents' lead; their greatest wish is to make their parents happy; and their strongest desire is to cooperate. Yet, this primal directive must be awakened and nurtured just like any other gift or ability. Rather than focusing on ways to manipulate children with fear and guilt, positive parenting focuses on ways to awaken children's willingness to cooperate. Using fear and guilt may effectively control children in the short run, but in the long run will weaken their willingness to cooperate.

WHY CHILDREN RESIST

When children resist a parent, it is often because they are wanting something else *and* they assume that if you just understood, you would want to support their want, wish, or need. Think about this for a moment. Most of the time as parents you are thinking, "What does my child want, wish, or need?" And then you take action to support your child. When children feel loved and supported, they will naturally assume you will change your request if you hear what they need, wish, or want. They assume you will adjust your request if you just understand the importance of their immediate wants and needs. Sometimes their resistance is just an attempt to communicate to you that there is something else they would prefer.

The power of understanding your children's resistance is that it immediately minimizes resistance. When children get the message that you understand what they want and how important it is to them, then their resistance level changes. It is not enough to just understand our children; we need successfully to communicate to them that we do understand. When children resist, the reason is that they mistakenly believe the parent doesn't understand what they wish, want, or need.

For example, a five-year-old child wants a cookie, but his mother wants him to wait until after dinner.

BOBBIE: Mommy, I want a cookie.

MOTHER: It's getting close to dinner time. I want you to wait until after dinner and then you can have one.

BOBBIE: But I want one now . . .

The child bursts into an angry tantrum. The mother first listens to understand Bobbie's resistance. After pausing to listen, she calmly says, "I know you want a cookie *now*. You are really angry because you want a cookie, and I won't give you one."

At this point, Bobbie relaxes for a moment and lets go of his resistance. This is because he thinks now that Mommy knows what he wants, he will get his cookie. Then the mother says, "You still have to wait until after dinner."

Sometimes this degree of understanding will be enough, but at other times a child will need more before he or she can cooperate. Most of the time children resist their parents simply because they don't feel heard or seen.

Children resist their parents simply because they don't feel heard or seen.

Let's explore what happens when Bobbie needs more time and understanding. After letting go of the anger that erupted during his tantrum, he now feels disappointed and sad. Although Bobbie continues to resist his mother, his resistance has a different quality. It has naturally shifted from anger to disappointment or sadness. Bobbie begins to cry saying, "I never get what I want. I don't want to wait."

Again, the mother gives understanding and identifies the unmet want, wish, or need. The mother says, "I understand you feel sad. You want a cookie and you don't want to wait. It's a long time to wait."

Clearly, the resistance is being minimized, but more importantly a deeper level of feeling is coming up. After a little crying, the child's fears will begin to surface. The child resists now by saying, "I'll never get my cookie. I never get what I want. I only want one. Why can't I just have a cookie?"

At this step, the mother avoids giving any explanation and continues to understand and identify the feelings and wants. She says, "I know you are afraid that you will never get your cookie. It is a long time to wait. I will make sure you get your cookie. I promise. Come here, sweetie, let me give you a hug. I love you so much."

At this point, Bobbie melts into his mother's arms and gets the love, reassurance, and support he really needed in the first place. Usually when children resist cooperating, they are needing something a little deeper. They need to be understood and loved and then they just need a hug.

TAKING TIME TO LISTEN

After reading this example of exploring with children the deeper feelings under their resistance, you may be thinking, "I can't do this every time my child resists me. You don't understand my children. They will consume all my time resisting." This would be true except that this skill really works. It minimizes resistance and creates more cooperation. If you take the time to listen and do it right, your children will resist less the next time and become much more cooperative.

Sometimes it does take an extra five minutes, but this is what your child needs. We spend *hours* each week driving around, shopping, and doing and getting things for our kids. While these external things are important, they are not nearly as important as supporting our children from the inside. Taking a few extra minutes to listen and identify our children's feelings, wants, wishes, and needs will not only give them what they really need, but it will also give parents more time for their own needs.

Taking the time to listen is much more important than getting to soccer practice on time.

Although using threats or disapproval may suppress children's resistance and save precious time, in the long run it creates in a greater resistance to other things. Quite often mothers complain, "But my child picks the worst times to resist. It seems that when I really have no time, then that is when they resist the most."

When children are not given permission to resist, frustration builds up inside and comes out when the parent is most distressed. This problem can be averted by taking time, when you do have the time, to listen to your children's resistance. Give them the message again and again that they are seen and heard.

If you never have time to listen, then you are not giving your children what they require. An ounce of prevention is worth a pound of cure. Don't wait until children's resistance builds up and then explodes. Take time to listen to their resistance whenever possible, and then it will not build up and come out when you really need them to cooperate and have no time or opportunity to listen and support them.

Try to remember that it is just an extra five minutes and that it is really worth it. Listening to your children is always more important than getting somewhere on time. When you take the time to listen to your children, they will automatically be more motivated to help you have time for yourself as well. Cooperation means you give and they give. Give the gift of understanding and your children will listen better and cooperate more.

THE TWO CONDITIONS

To communicate that you hear or understand a child's pressing needs, wishes, and wants, two conditions must be met. The parent must communicate the validating message, but the child must also be aware of the need to be heard and not just his or her desire for a cookie now. By setting a boundary (that is, "You can't have cookie now"), children feel their resistance; they are not aware of their underlying need to be heard.

The next step is for the parent to identify their child's emotion of anger or frustration in a calm and warm way. When a parent acknowledges this emotion, children become aware of what they feel. Although children may be angry, they are not yet aware that they feel angry.

With an awareness of their feelings, another doorway opens for the child. At this point, children can also identify and feel their need to be heard. When a child feels the need to be understood and that need is fulfilled, then the biggest part of the struggle is over. The child recognizes that he or she is being heard. This feeling of being heard is then confirmed when the parent says what the child is wanting.

All this occurs in a brief moment when the parent says, "I know you want a cookie *now*. You are really angry because you want a cookie and I won't give you one." The child's response is a complete yes. It is hard to keep resisting when you are feeling yes, and you are being heard and understood.

The cookie example worked because both conditions were met. The parent communicated her understanding and the child felt his need to be understood being met. Though this technique is most effective with sensitive children, it works with all the four temperaments. It may just take a little longer with the sensitive children, because they need the understanding so much.

The more sensitive a child is, the more he or she may need to go deeper into their feelings. A parent can guide the child to deeper levels by simply remembering this easy format of emotions. Under children's resistance are first anger, then sadness, and then fear. By giving children a chance to go deeper and feel these feelings, a door in their heart opens and they can feel their real and most important needs being met. Unless children go a little deeper, they stay on the surface only resisting and wanting the cookie.

Under children's resistance are first anger, then sadness, and then fear.

For sensitive children, parents need to focus primarily on drawing out the anger, sadness, and fear, while acknowledging that they clearly understand what their child wants.

For active children, parents need to focus on a few of the primary feelings, but acknowledge what the child is doing or wanting to do. For example, you might say, "I know you have stopped everything to come over here and get a cookie. You are really angry because you want a cookie and I want you to wait until after dinner." If a child is active, you can succeed in giving better understanding by just elaborating a little on what is physically happening or not happening and by letting the child know directly what you want him to do.

For responsive children who need redirection, you could add a little phrase like this, "I know you want a cookie *now*. You are really angry because you want a cookie and I won't give you one. Let's go over here and wrap up this cookie for you to have after dinner. Tonight we are going to have pink salmon and fluffy white potatoes. Look at these potatoes . . ."

For receptive children who need more rhythm, add the element of time and it will work a little better. Use the phrase, "I know you want a cookie *now*. You are really angry because you don't want to wait. Right now it is time to get ready for dinner and after dinner it will be time to eat dessert. First we eat and then we have dessert." Receptive children need a little rhythm and then they can relax.

Each of these four different approaches works best when applied to the appropriate child, but the original example would also work. Remember that every child has a little bit of each of the temperaments. Any of these approaches will work.

HARD-LOVE PARENTING

When it comes to dealing with our children's resistance, there are generally two different approaches: soft love and hard love. Hard-love parents mistakenly believe, "If I tolerate my children's resistance, then I will spoil them. They must always remember who is boss." Although this limited thinking is now out of date, it is still partially true. To have a healthy sense of security in life, children need to always remember that the parent is the boss.

Although children may love being the boss, it works against their well being. Children need to play in the magical world of childhood without the burden of being responsible. Too many choices will create an inner insecurity that gives rise to a host of problems. A child will disconnect with his or her natural willingness to cooperate and become demanding, selfish, needy, or just more resistant. An updated adjustment to the old adage, "Spare the rod and spoil the child" is "When a child forgets who's boss, you spoil the child." The new message we need to give our children is that it's okay to resist, but remember mom and dad are the bosses.

It's time to update and adjust the old adage,
"Spare the rod and spoil the child."

The wisdom of the past must always be updated with new adjustments. To create order in society, we no longer need to take adulterers and stone them outside the city walls. In a similar way, we don't need to spank our children or be intolerant of their resistance. Hard-love approaches must be rethought and adapted to meet the new needs of every generation.

The hard-love approach teaches children who is boss, but does not tolerate children's natural resistance. While fear- and guilt-based approaches used to work, they now create their own set of problems. As we have already discussed, children do not need to be beaten and punished to create a willingness to cooperate. Children are born already willing to cooperate, but if they are not permitted to resist, they will either be weak and obedient, or they will attempt to find their inner power through rebelling.

Punishment may make them obedient in the short term, but later on they will rebel. Children today are rebelling earlier and earlier. This rebellion not only makes parenting more time consuming, difficult, and painful, but it obstructs a child's natural development.

Some experts today will say it is good for your child to rebel at puberty, that it is normal for a child in puberty to stop talking to his or her parents or looking to them for love and support. Although things do change at puberty, it doesn't mean a child has to rebel against the parents or stop going to them for support. The huge disconnection between parents and teenagers is not normal or healthy—it is just common.

The huge disconnection that is occurring
today between parents and teenagers is not
healthy—it is just common.

Though teenagers naturally feel a greater need for peer support, this does not mean they no longer have a need for their parents' guidance and love as well. It is not a given that a teenager will defy or rebel against their parents. Yes, it is a time for them to explore their individuality, but this does not mean that they will rebel or disconnect from a healthy willingness to cooperate, please, and follow the direction of their parents.

To live a fulfilled life today, it is not enough to surrender your will to the rules and live obediently under the rule of the boss. It doesn't help our children to break their will and teach them to follow rules mindlessly and heartlessly. Children today have the potential to create the lives they want.

Our children have the power to make their dreams come true, but this power must be nurtured. It is a creative power. When there is a problem or obstacle, a creative child or adult does not just accept and give in. Instead, creative people look for another way, a way to get what they want and to serve the needs of others as well. By awakening the spirit of cooperation in our children, this kind of creative intelligence is awakened. By raising our children to simply be obedient, we fail to give them the winning edge they need to compete and succeed in the world today.

In raising merely obedient children, we fail to
give them the winning edge.

Success in life doesn't come from following rules; it comes from thinking for oneself and following one's heart and inner will. This natural ability is first nurtured by strengthening the child's willingness to cooperate. Demanding obedience from your children numbs their inner will. It closes their mind and heart and disconnects them from their potential to create the life they are here to live. When children get the message that it is okay to resist, but remember mom and dad are the bosses, they have the opportunity to keep their mind and heart open and nurture the ability to know their own will and wish in life.

Success in life doesn't come from following
rules, it comes from thinking for oneself and
following one's heart and inner will.

When parents are able to respond to a child's resistance calmly, without threats of punishment or disapproval, then a child gradually learns how to deal with the resistance she experiences in the world. When confronted with someone who is not willing to cooperate, she knows how to deal with the situation without mindlessly giving in or demanding that the other person give in.

Positive parenting teaches children to navigate through life's obstacles with understanding and great negotiation skills. They know with certainty the power of listening to minimize resistance and increase a person's willingness to cooperate. What was done to them, they do to others. When parents listen more to their children, their children automatically learn how to listen as well.

SOFT-LOVE PARENTING

Many parents have given up hard parenting. They recognize the importance of listening, but don't understand the importance of being the boss. They seek to avoid their children's resistance by listening and then placating the child. They listen, but then cave in to their child's resistance to make the child happy. They cannot bear to see their children unhappy, and so they make whatever sacrifice they can to please them. This brand of soft-love parenting does not work and has made many parents suspicious of new nurturing skills of positive parenting. Fortunately, positive-parenting skills work right away. They work in the short run and in the long run.

The failure of soft love parenting make many parents suspicious of positive-parenting techniques.

Soft-love parents sometimes give in to their child's wants and wishes, because they just don't know what else to do to stop the tantrum. They refuse to do what was done to them when they were growing up, but they don't know another way that works. They know that spanking and shaming doesn't work, but don't know what does. By indulging their children, they mistakenly give the message that throwing tantrums or being demanding is a good way to get what you want.

Soft-love parenting tries to please and placate the child.

Soft-love parents will do whatever they can to avoid a confrontation with their child. They don't know what to do when the children resist a request and develop new ways to avoid their children's resistance and to motivate cooperation. They give the message that it is okay to resist, but they don't establish that they are the boss.

To avoid a child's resistance, many parents are even taught by well-meaning experts always to give a child a choice. Giving choices will lessen resistance, but it doesn't create cooperation. It is another way of giving a child too much power and weakening your power as boss.

> Giving choices will lessen resistance,
> but it will not create cooperation.

Until the age of nine, a child doesn't need choices. Having too many choices pushes a child to grow up too soon. One of the greatest sources of stress for adults today is too many choices. Directly asking a young child what she wants puts too much pressure on the child. Always asking children what they want or how they feel weakens a parent's ability to maintain control.

Greater freedom and responsibility creates more anxiety unless we are ready for it. Children younger than nine are not ready for it. They need strong parents who know what is best for them, but who are also open to hearing their resistance and discovering their wants and wishes. After discovering a child's wishes and wants, parents can then decide to change their direction or hold strong. Either way, the parent is still in charge.

This concept is similar to our court system. Once a case is tried, it cannot be reopened unless there is new evidence.

In a similar way, although a child can resist, and it doesn't mean the parent will budge from his or her point of view. If the parent gets new insight by listening to the child, it is fine to reconsider what is to be done. Parents may change their mind, not because they are afraid of resistance, but because new information has been considered that has changed their point of view. Like the court system, parents do not change their request unless new information becomes available.

**Children need strong parents who know
what is best; they don't need more choices.**

Soft-love parents don't know that resistance is an important need that children have. Children need to test the limits and make sure what you want them to do is really important. Otherwise, they have things to do that they consider more important. Just as children need permission to resist and test the limits imposed on them, they need a strong parent who will listen and then decide what is best.

Positive parents always decide, because they are the boss. Children are not ready to be self-employed. They need a boss. Without a boss, they begin to self-destruct. Permissive, soft parenting minimizes resistance in the short run, but weakens children's willingness to cooperate. As a result of hard parenting, girls tend to lack confidence, while boys lack compassion. As a result of soft parenting, girls tend to have low self-esteem and, later in life, give too much, while boys become hyperactive and lack confidence and discipline.

**Children are not ready to be self-employed;
they need a boss.**

Using positive-parenting skills means hearing your child's resistance and then deciding what is best. Deciding what is best doesn't mean that you do not deviate from your original position. As your children develop a greater awareness of what they need and want, often they become great negotiators and are able to persuade you to change your mind.

There is a world of difference between giving in to your child's feelings or wishes and changing what you think should be done. Parents are the boss, but they must not always rigidly hold on to their request or point of view. To listen to a child's resistance means to consider what he or she is feeling and wanting and to decide what is best and then to persist.

It is fine to begin asking your nine year old and older children what they feel, want, wish, or need directly. Then around the age of twelve to fourteen, it is time to begin asking teens what they think. The development of abstract thinking at puberty signals children's ability to start making decisions for themselves. How we communicate with our children always needs to be age appropriate.

At every age, children need a clear message that it is okay to make mistakes. The best way to teach this is by learning from your own mistakes. The truth is parents are not always right, and they don't always know what is best. They can know better if they consider and hear their child's resistance. If parents change their point of view, it should be because they learned something and think a change is best. They should not change their point of view to minimize a child's resistance. Placating your children to minimize resistance only paves the way to greater resistance in the future.

LEARNING TO DELAY GRATIFICATION

Whether parents use hard love or soft love, their children do not get an opportunity to experience and cope with their resistance to limits. Expressing resistance not only defines the limits of a child's space and influence, but also helps the child adjust. Learning to accept the boundaries and limits of time and space is a big lesson in life. Pushing up against life's limits can teach children to embrace those limits without having to deny themselves. One great benefit of expressing and then letting go of resistance is the ability to delay gratification.

Many studies have shown that children who are able to postpone gratification are more successful in life. This really doesn't require a study to realize. Look around you and you will see that people who succeed are people who patiently persist in achieving their goals. They do not throw in the towel when they don't get what they want right away. They don't lose touch with or deny their wants and wishes just because life doesn't give them what they want when they want it. They bounce back from life's setbacks with renewed energy and enthusiasm.

The ability to delay gratification is also the ability to be happy and at peace even though you don't have everything you want. When children are able to resist their parents and then gradually let go of their resistance, they are learning to accept what has to be. They are accepting limits in a spirit of cooperation and trust that everything is and will be okay. Ironically, it is the ability to express resistance that allows us to become more fluid in life. A clear acceptance of what has to be allows us to see more clearly what can be changed. This not only brings greater peace but the motivation to persist in trying to change what can be changed.

By expressing and then letting go of
resistance, children learn to accept what
has to be.

Within every person is the natural ability to bounce back from life's limits and curve balls. When children resist not getting what they want, and parents can identify and understand the feelings underlying the resistance and can communicate that understanding to them, then the children discover their natural ability to be happy and accepting even though they are not getting what they want. By giving children the loving understanding they need, having what they want right away becomes no longer so important.

When children are demanding about their wants, it is usually because they are not getting what they really need. Likewise, when adults are unhappy because they are not getting what they want, it is because they are really not getting the love and support they need in their lives. The love we need is always available, but we are not seeing it.

Children need boundaries to push up against. When they don't get them, they are restless and insecure. When they get their way too often, then what they get is never enough. It is only when we are feeling our needs that we can appreciate what we get. Resistance reconnects us to what we need instead of being focused too much on what we want.

When parents listen to children's resistance and appropriately assist them in being aware of their feelings, wants, wishes, and needs, the children learn to be more aware of what is most important and is not tossed around so much by the ups and downs of life. Most adults today suffer from headaches, heartache, stress, distress, backaches, and other illnesses because they have focused too much on what they want and not enough on what they need.

MEETING YOUR CHILDREN'S NEEDS

You can't always give your children what they want, but you can give them what they really need. If you don't focus on providing what they need, then you and your children suffer increasing resistance. Contained within a child's resistance is the greater need to be seen and heard, cared for and loved. To know themselves, children are completely dependent on how much their parents see and hear them. When a child resists getting ready for school, refuses to eat vegetables, or simply doesn't listen to your requests, it is a clear message that the child needs more time, attention, understanding, and direction. Your children need you to know what they need and for you to provide it.

In many cases, just the act of listening to a child's feelings or resistance will give a child what he or she needs. But if a child needs something more, like more structure, then just listening will work only temporarily. Let's have a look at an example.

A mother might ask her two sons, ages six and nine, to stop fighting. After hearing their resistance and understanding their frustration, the mother is able to bring the children back to feeling more cooperative. However, if those children don't have enough structure within five or ten minutes, they will begin fighting again.

In this example, besides feeling heard, the children also need some kind of structured activity with rules, otherwise they will require more supervision. In this example, listening to the child is not enough.

**When children don't know what to do they
often forget how you want them to behave.**

When you just can't give your children what they need in the moment, there is another way to get the cooperation you want. These new skills to motivate your children work, but they still don't replace giving your children what they need. Though they will create cooperation and motivation, they won't give children what they need to develop. These skills will get your children to do what you want, when you want it, but your children will still have other needs such as understanding, structure, direction, and rhythm. Just as punishment was used as a deterrent in fear-based parenting, rewards are used to motivate in love-based parenting. In the next chapter, we will explore how to motivate your children.

6

News Skills
for Increasing
Motivation

In the past, children have been controlled or motivated to behave primarily by the threat of punishment. When a child starts to misbehave or is uncooperative, most parents' gut instinct is to threaten the child. We say or feel things like, "If you don't listen, you'll really be in trouble" or "If you don't stop crying, I will give you something really to cry about." With little children, we might raise our hand to spank them or give a certain look that means if they don't cooperate they will be punished. By using punishment the threat of loss, violence, pain, or increased suffering is employed as a deterrent.

Using fear as a deterrent appears to work, but it doesn't awaken children's natural motivation to cooperate and to help a parent. As noted earlier, obedience and cooperation are very different. Children need to be a willing helper to be truly cooperative and breaking a child's will with punishment is not the answer. It is difficult to let go of punishment, because it works so well in the short term. Although we don't want to punish our children, we just don't know another way. Punishment seems inhumane, but without it, our children become spoiled, demanding, disruptive disrespectful, or unmanageable.

Most parents don't want to punish their
children; they just don't know another way
that works.

Responding to this need for change, some experts suggest "giving children consequences" to their behavior. For example, they take away something as a consequence of misbehaving and instead of calling it a punishment, it is called a consequence. This is an attempt to take the shame out of punishing. Instead of giving the message, "You are bad so you should be punished," the child gets a more positive message: "It's okay that you made a mistake, but now you will learn the consequence of your behavior." Though this technique lessens guilt and is more humane it is still based on fear. This approach is much better than punishing, but it doesn't awaken children's natural desire to cooperate. In a sense, it is a nicer way of saying, "you will be punished."

A SHORT UPDATE ON PUNISHMENT

Our entire history in the last five thousand years is based on punishment as the model for control and rehabilitation: an eye for eye; justice for the victim; make the abuser pay for the crime. In the past, avengers have felt satisfied, but today this satisfaction is merely a temporary relief and the pain of the victim lives on.

Even practically speaking our system of punishment is not working. Our penal system costs taxpayers $25,000 a year per prisoner. To give one twenty year old a twenty-five-year sentence in jail will eventually cost taxpayers $625,000. This money certainly doesn't pay back any debt. Think for a moment how much better that money could be used to prevent the crimes in the first place.

To punish one person with a twenty-five-year jail sentence will cost taxpayers $625,000.

So much is still based on this out-of-date notion of the efficacy of punishment. Today, if you follow the precept "an eye for an eye," eventually everyone in the world will be blind. Although we know in our hearts that punishment is out of date, a clear alternative has not yet been realized or discovered.

If you follow the precept an "eye for an eye," eventually everyone in the world will be blind.

Today, although rules are important, punishment is not. One day far in the future as consciousness continues to grow, even rules will not be important. In the past, punishment was important because people were not yet capable of knowing what was right within their own hearts and minds; making sacrifices to their Gods and punishment to the wicked was the only way to motivate them.

WHY AND WHEN PUNISHMENT WORKED

Inhumane punishment directly caused pain (cutting off a finger or whipping or stoning someone) while more civilized punishment would take away money (fines) or freedom (prison). In a more humane manner, civilized punishment caused people to feel the pain of loss. To be a better person and avoid making mistakes, many people worshipped God by making offerings or sacrifices. By giving up something for God, they felt the pain of loss, and as a result they became more aware of right and wrong or the best course of action for them. Making sacrifices to God had clear benefits.

By feeling pain we are automatically induced to correct our thoughts and actions.

Though this may seem bizarre, think for a moment about your common experience. Often, after a loss, when we feel the pain of that loss, we experience our regret and resolve to do things differently and learn from our mistakes. Feeling pain motivates change to avoid pain in the future. In addition, by being more aware of our feelings, we can tap into greater creativity and intuition from our inner potential. The ability to know right from wrong comes from our feelings. Feelings, whether negative (pain) or positive (pleasure), help us to make needed adjustments.

With this motivation, we open our minds and question what we have done. This inner questioning is the basis of self-correction. Unless we are motivated to change, we remain stuck in our narrow and limited ways of thinking. Pain is our greatest teacher, because it motivates us to make adjustments in the way we do things. It causes us to question and rethink what is best for us and for others.

When people were numb to their feelings thousands of years ago, they needed punishment to connect with their feelings. With a greater awareness of their feelings, they could then accept or recognize what was right and reject what was wrong. Gradually, after being punished for centuries, just the thought of punishment was enough to awaken the feelings. Rule by punishment was necessary to sustain order and to enable a person to lead a good life.

THE POSITIVE SIDE OF PUNISHMENT

Induced pain, whether through punishment or through spiritual sacrifice, awakened people to their feelings and increased their limited awareness of what is right and wrong. In this manner, punishment was a tool or skill to induce the feeling of pain, and, in various degrees, it motivated change.

Christian monks, even into this century, would often punish themselves to become more holy. They would whip themselves as a daily practice to deepen their connection with God. As extreme as this may sound, self-flagellation was widely practiced. This and other forms of self-mortification are still practiced; it is not uncommon to give up comforts and pleasure in the name of being spiritual.

These practices are no longer necessary. The time to give up our lives for God is over. It is now time to live our lives fully for God. Everyone deserves abundance, prosperity, success, health, and love. We do not need to deprive ourselves of life's pleasures to lead a life of goodness. Likewise, we do not need to deprive our children. If we want our children to live a life of abundance, we must find another way to motivate them, otherwise, after making mistakes, they will tend to punish and deprive themselves as well.

The time to give up our lives for God is over.
It's now time to live our lives fully for God.

By feeling our loss naturally, we experience a greater awareness of what is right or wrong, and we feel the motivation to change. Another more biblical way of saying that we have "a greater awareness of what is right and wrong" is to say that we have "a greater awareness of the will of God."

Today we no longer need to punish ourselves or our children to act in accord or in cooperation with the will of God. We are born with the ability to know and do what is right, but having an ability or talent isn't enough. To be realized, this ability must be nurtured and developed.

Children today have new needs. By fulfilling these needs, we directly nurture their ability to cooperate and increase motivation to yield their will to their parents. Our children today do not require outdated punishment; they have a greater potential and require a new and different kind of support.

Our children today have a greater
potential and require a new and different
kind of support.

This new ability has been in the making for a long time. Two thousand years ago, Jesus taught this simple message: *As you open your heart to God, your self, and your neighbor you will know the will of God; in the silence of your heart, a quiet voice will speak to you. By looking within, you will find, right now, the heaven you seek.*

This quiet voice, often referred to in religious texts, comes from your heart and mind being open. It comes from feeling. As parents are able to speak and act more from love, their children also learn to listen, not only to their parents, but to the feelings of love in their hearts. Then they are motivated, not out of fear, but by love.

When parents raise children with open hearts, minds, and

strong free wills, this quiet voice is not some exalted experience that only saints can hear, but a common experience motivating children's daily behavior. When we can look within or "feel" then we discover that the kingdom of heaven is at hand—it is here and now. When we live from our hearts, then we have succeeded in bringing heaven to earth.

After two thousand years of trying to understand how to be loving and get what we want and need, we have finally arrived at our goal. It is now possible to apply the principles of love to raise our children. Even if these new skills for positive parenting were available thousands of years ago, they would not have worked. They would not have worked for everyone even fifty years ago. A shift in our global consciousness has made it possible for these new skills to work. Now the old skills do not work.

THE SIMPLE PROOF

Our prison system has proven that in a free society punishment no longer works. Under a dictatorship, the threat of punishment is extreme and fear is everywhere; that is how a dictatorship maintains order and low crime. In a free society, punishment has failed. Today, instead of building more schools, we are building more prisons. In many parts of the country, when someone is punished in our prison system, they often emerge, not rehabilitated, but as better criminals. Rehabilitation centers could better be called criminal training centers. Clearly, in a society that allows personal freedom and respects human rights, the old ways of maintaining order by means of punishment are out of date. We cannot preach love and then turn around and punish the weakest elements of society. Fortunately, some prisons today are focusing more on methods to rehabilitate and not just on punishment.

Punishment doesn't work in a free society, and it doesn't work in loving families. The more children feel nurtured and loved, the more confusing punishment is. We cannot nurture our children and open their minds and hearts to be strong, creative, and capable, only to turn around and threaten them like animals. We cannot seek to make them feel good about themselves and then make them feel bad when they make mistakes.

We open our children to feeling good about themselves and then turn around and make them feel bad.

It is more damaging to open up children and then punish them than to ignore their feelings and wants, and occasionally punishing them to maintain control. If we are to give our children the opportunity to open up their minds and hearts and develop a strong will, we must learn another way to motivate other than punishment.

Even animal trainers are learning news ways to train dogs, horses, tigers, and other animals without punishment. I learned more about parenting by talking with animal trainers than from many of the parenting books available to parents. There is so much confusion when it comes to parenting, and one of the most controversial issues is punishment.

Everyone senses that punishment doesn't work and is inhumane, but they don't know a different way. Many are resistant to the idea of giving up punishment, because the soft-love type of parenting clearly has failed. Children who are not punished are often unruly, undisciplined, and disrespectful to each other and to adults and teachers. Yet every parent, at some time, has felt in their quiet moments that there must be a different way. Fortunately, there is an alternative to punishment, and the collective consciousness of our planet is ready for it to work.

THE ALTERNATIVE TO PUNISHMENT IS REWARD

Instead of motivating children with punishment, children today need to be motivated with rewards. Instead of focusing on the consequences of negative behavior, positive parenting focuses on the consequences of positive behavior. Instead of using a negative outcome to motivate children, it uses a positive outcome.

There is no greater motivator, other than children's inner desire to cooperate, than their desire for reward. Many times it is the outer reward or acknowledgment of success that awakens children's inner desire to cooperate. Every child wants special time with his parents. Every child gets excited when you mention dessert. Every child loves presents. Every child looks forward to a party or celebration. Every parent has noticed how warm, friendly, and cooperative children are when they want something and think they can get it.

Getting "more" or the anticipation of getting more awakens something inside, and a child jumps up with a big yes. The expectation of reward gives children the energy and focus to respond to their parents' need for cooperation and help. The promise of more inspires everyone, old or young, to cooperate. Rewarding, rather than punishing, your children will increase their willingness to cooperate.

The promise of more inspires everyone,
old or young, to cooperate.

While parents are sometimes slow to adopt new ideas, successful businesses are not. To survive and flourish, businesses must adapt to change very quickly or they will get left behind. The airlines, for example, clearly know that giving perks, incentives, miles, and extra miles is the way to motivate people to fly their friendly skies. Most successful companies now routinely offer special rewards for employees who excel.

Using incentives in the business world is common sense, but when it comes to parenting, there is a strong undercurrent of belief that rewarding children is like bribing them, and, if you need to bribe them, you really aren't the boss. For some, motivating your children with a reward somehow implies that you as a parent are weak and your children are running the show. Yet, those who promote this belief will turn around and punish their children to make them behave . . . and a punishment is just a negative bribe.

This message is hard to hear for those who have gone against their heart's instincts by using their heads to justify punishing. Many parents even say as they spank their children, "This hurts me more than it does you." Their hearts were speaking, but their minds were not yet ready to listen. They love their kids, but just did the best they knew.

Already thousands of children have been successfully raised without punishment or using threats to keep control. Their parents didn't punish, and it worked fine. These children were not unruly or undisciplined in their behavior and turned out great. Yet, on the other hand, millions of parents have clearly failed by using soft love, hard love, or by going go back and forth.

THE TWO REASONS A CHILD MISBEHAVES

To understand why the conscious use of reward works best, we need first to explore the two reasons children misbehave. The first and most important reason children today misbehave is that they are getting what they need to stay in touch with inner feelings. Remember, nonfeeling children need punishment to reconnect with their feelings. Children today just need understanding, structure, direction, and rhythm, and they will automatically be more in touch with their feelings.

Children go out of control when they are not
getting what they need.

When children don't get what they need, they go out of your control and misbehave. They misbehave not because they are bad, but because they are out of your control. When children are getting what they need, they remain under your control and cooperate. You may have a great car that works perfectly, but if you let go of the steering wheel it will quickly crash. Unless parents keep control, their children will crash.

The second reason a child misbehaves is determined by how the parent then deals with the child's unruly behavior. By continuing to focus on the negative behavior, children will continue to behave in a negative manner. When you focus on negative behavior, that is what you will get more of. Punishing children forces them to focus on a negative behavior rather than focusing on the positive.

WHY GIVING REWARDS WORKS

Rewarding your children for positive behavior means focusing on the good that they do. Punishing your children focuses your attention on the bad that they did and reinforces the old idea that they are born bad and need to be rehabilitated. By focusing on the bad, the good does not have the opportunity to come up and be expressed.

What you put your attention on grows. When you punish a child, a lot of attention gets put on a child's negative behavior. A parent might even say, "I'll teach you a lesson that you will never forget." The opposite of punishment is a forgiving attitude that clearly states it is okay to make mistakes and then forget about it and move on. What is more important with children today is nurturing their needs and directing them in ways to make them successful.

If you reward a child's positive behavior, that is what will increase. Rather than look for and focus on a child's mistakes, parents need to try "catching" the child doing things right. Whenever your children are moving in the right direction, acknowledge their success, and they will continue to move in that direction.

Rather than look for and focus on children's mistakes, try "catching" them doing things right.

For a young child, aged four to nine, make a chart of a few chores and positive behaviors. Before bedtime, review the list and stick on stars or bright and colorful stickers next to any chores completed for that day. If they didn't do the chore or positive behavior, just leave a blank and don't put much attention on it. Have a neutral to bored attitude regarding the blanks and focus enthusiasm and positive feelings regarding successes. Each star can mean a point, and when the points add up to twenty-five, then you should do something special such as have twice as much reading time or go to a baseball game. This activity then becomes another special memory linking the child back to feeling acknowledged and successful.

Have a neutral to bored attitude regarding mistakes and focus enthusiasm and positive feelings regarding their successes.

Keeping a chart helps parents remember to acknowledge whenever children happen to do the right thing. Most parents are not even aware of how much they verbalize what their child is doing wrong. With this insight, it becomes more clear why children don't listen. If we drown our children in negative statements, we cannot expect our children to cooperate. This is a list of thirty-three common expressions to assist you in becoming aware of things you may be saying.

NEGATIVE ACKNOWLEDGMENTS

You didn't put your books away.

Something is wrong with you.

You are being too loud.

Don't be mean to your sister.

Your room is a mess.

How many times have you forgotten your jacket?

When are you going to grow up?

You are not listening to me.

Don't go over there.

Don't play with your food.

I wish you had been a boy.

Stop daydreaming and look at what you are doing.

Stop running around.

You are playing too rough.

You are being bossy again.

No one will like you if you behave that way.

You didn't say thank you.

You didn't say please.

Keep your mouth closed when you chew.

You didn't do anything I asked.

You are watching too much TV.

Turn down your music; it's giving me a headache.

Stop whining.

You can't do anything right.

Try to remember this time.

Slow down, you are going too fast.

You are not fun to play with.

Don't be stupid.

You are being a big baby.

I can't deal with you.

There's no way you can do that.

That doesn't make sense.

This is all your fault.

By being aware of how often we give negative acknowledgments, we can begin to stop. Instead of dwelling on the problem or punishing our children for their imperfections, we can begin asking our children to be the solution by directing them. If we can't say something positive or direct our children in a positive way, then we shouldn't say anything. These are some examples of directing a child rather than focusing on the problem and then punishing.

Dwelling on the Negative	*How to Give Positive Direction*
You are not listening.	Please give me your full attention.
I can't deal with you. I need you to . . .	Please I want you to cooperate.
Look at the way you are dressed.	Would you go put on that new blue shirt; it would look great with those pants.
There's no way you can do that.	Let's see if there is another way to do this.
Don't be stupid.	Let's go over this one more time in greater detail.
Slow down, you are going too fast.	Would you please slow down?
You didn't put your books away.	Would you put your books away please?
Don't sing at the dinner table.	Please don't sing at the dinner table.

| Stop whining. | I don't want to talk about it anymore. |
| You are being selfish again. | Please I want you to remember your manners. |

Certainly, we have to correct our children, but, instead of focusing on their behavior in a negative manner, we can give them a chance to change their behavior for the better. Even correcting our children's mistakes in a positive manner becomes counterproductive. We need to acknowledge them three times more for their positive behavior. It takes more positive to balance the negative. Children often stop listening to their parents because they are not getting enough acknowledgments of the good.

Here are thirty-three examples of catching your children doing something good or right and letting them know:

CATCHING YOUR CHILD BEING GOOD OR DOING THE RIGHT THING

You put your book away.

Everything looks so nice and orderly in your room.

You are so smart.

That was so kind of you.

I appreciate your using your inside voice.

You did such a good job.

You are being so helpful.

Everything is going so smoothly.

You remembered to use your manners.

You are such a big help to me.

It is so much fun playing with you.

I love you and I love being your Mom/Dad.

That was a good shot.

Thank you for listening and not interrupting.

You followed all my instructions—good job.

You are using your silverware so nicely tonight.

You are a hard worker.

You stayed right on track.

You are such a great helper.

This is a wonderful picture, I love it.

Look at what you have done—it is great.

It's okay, I know you always do your best.

You are really remembering your table manners.

You are being so cooperative tonight.

I noticed that you shared your toy; that was very considerate.

You got dressed all by yourself.

You did that all by yourself.

It's good to ask for help and you came and found me.

You did a terrific job.

You cooperated the whole time, thank you.

You are so loving with animals.

Thanks for helping, I know I can depend on you.

You are looking so bright and healthy today.

By pointing out and acknowledging positive things about your children and their behavior, they will see themselves as successful and good. This positive image of themselves will not only motivate them to cooperate but will also create self-esteem, confidence, and a sense of competence.

Soft-love approaches generally endorse this concept of giving lots of positive acknowledgment. When children of self-love parents feel insecure and experience low self-esteem, some experts mistakenly assume that giving positive acknowledgment doesn't work.

It does work. What doesn't work with soft love, as we have already explored, is not facing children's resistance to doing what the parent wants. The soft-love parents dread confrontation and regularly cave in to children's demands to avoid having to deal with a tantrum. Placating, not positive acknowledgment, spoils the child.

THE MAGIC OF REWARDS

Positive parenting focuses on motivating children to cooperate in a variety of ways. We use asking and not ordering. We focus on nurturing children's needs rather than trying to "fix" them. We listen to resistance and don't lecture or get upset. When that doesn't work, we use rewards to motivate children to cooperate. If a parent only uses rewards, then it is counterproductive. Rewards motivate, but they do not give children all of the understanding, structure, direction, and rhythm they need.

Rewards are particularly useful at those times when we don't have the time or opportunity to give children what they need. There are times when we have not been able to meet all our children's needs or we don't have time to meet what they need in the current situation. At those times using rewards will temporarily create the cooperation you need. Following are some common examples of situations that cause our children to not cooperate because they are not getting what they need.

WHY CHILDREN RESIST OUR DIRECTION

- They are disappointed and need to talk for a while and be understood and you don't have the time to listen.

- They are tired and need a nap; their natural rhythm has been disturbed.

- They are hungry and need to be fed.

- They don't know what to expect and need more time to prepare.

- They were not prepared with the bigger picture of what you expect and what the rules are.

- They are overstimulated from too much TV, too long of a shopping trip, too many people, too much fun, too much dessert, or simply too many activities.

- Something else is bothering them and they need to talk it out or get help. They could have an earache or someone may have been mean to them that day.

Sometimes outside influences and stress beyond your control that disturb children and create resistance. For example: You are in the grocery store or on an airplane and your child is being affected by the stress of others who don't want to hear a child cry.

Remember, children are supposed to throw tantrums and express their resistance in order to get the understanding that they need to define themselves. If children haven't had enough tantrums at home because the parents have coddled them too much, they tend to have tantrums in public when the parent's can't coddle them. They are used to being placated. In a public situation or stressful situation when a parent can't give more, they become demanding and throw a tantrum.

When children resist it could be because any of these reasons listed above and many more. If children resist, then in some way, they are not getting all their needs met. We do not live in a perfect world, and as parents we are not perfect. We cannot always give our children what they need no matter how much we know or have to give. Occasionally, our children will resist our direction when we just don't have the time or resources to give them the attention, understanding, structure, redirection, or rhythm they need.

Resistance is inevitable because parents are not perfect and cannot always give children what they need.

Rather than mistakenly assuming that our children don't want to cooperate, we need to realize that they don't have what they need to cooperate. If a car doesn't run because it is out of gas, it is not appropriate to assume that the car is resisting you or broken in some way. When children resist, they are unable in that moment to cooperate; they don't have what they need to reconnect with their inner desire to cooperate. The purpose of rewarding children is to give them a little more fuel to connect with the part of them that wants to cooperate.

> The purpose of rewarding children is to awaken the part of them that wants to cooperate.

Instead of trying to get control with threats of punishing or spanking, at those times when our children resist cooperating, we can regain cooperation by means of rewards. Giving a reward will often evoke cooperation.

UNDERSTANDING REWARDS

Imagine you were asked to work overtime and you automatically felt resistant. Then you were informed you would get paid twice as much for each hour of overtime. Immediately you would become more cooperative. Just as the promise of more will motivate you, it also works, perhaps even better, with your children. It is natural. Let's look at a few examples.

When a child refuses to brush her teeth say, "If you go and brush your teeth now, we will have time to read three stories instead of just one."

I still remember when I started consciously using rewards with my children. One of my children consistently resisted brushing her teeth before bed. Nothing would work. Then, after taking a parenting class that recommended rewards, I used this one simple phrase and it worked. I was amazed. Just by letting her know that we would have more time to read, she jumped up to brush right away without any fuss. This one simple shift brought immediate results and changed my whole parenting approach.

With the support of giving small rewards, the job of parenting becomes much easier. In many cases, a child's resistance just melts away with a reward. With occasional rewards, a child is reconnected with her natural desire to please the parent, and she automatically cooperates more of the time.

Giving small rewards makes parenting
so much easier.

Yet, some parents worry that their child may take advantage of this kind of support and always demand a reward before they do anything. Fortunately, this does not occur. When used with the other skills of positive parenting, giving rewards actually awakens and strengthens a child's willingness to cooperate without rewards. Once a child has been motivated to do a particualr behavior with rewards, soon after she no longer requires the reward.

When children are in control, they don't need rewards. They only need rewards to help them come back into your control. Rewards are only needed at times when chi;ldren are out of control and disconnected from their natural desire to please their parent. Once a particular behavior is established, then the child doesn't require a reward to continue doing it. Giving the reward of three stories before bedtime does not make a child demand a reward for cooperating at other times.

Until I experienced the power of giving rewards, I resisted giving rewards, because I thought it was like bribing. When it worked so well, I had to begin considering its merits and rethink why I resisted giving rewards. When one of my children resisted my direction, my gut reaction was to make a threat. This was how my father parented me, and so, at times of frustration, my reaction was to threaten as well. As soon as I discovered a better alternative, punishment and the threat of punishment became a thing of the past.

My new challenge was to find appropriate ways to give rewards. The reward must be linked in some way to the behavior the child is being asked to modify. Ideally, a reward is the natural consequence of cooperation. If the child brushes her teeth right away there really is more time to read stories before bed. When a child is resisting putting on her coat, the natural consequence of getting to school sooner may not seem like a reward. In some cases, however, it might work. You could say, "If you put on you jacket now, then I will have time to look at your paintings at school."

There is one reward that works all the time, and you don't have to think too hard. It is the gift of time. You can say, "If you cooperate with me now, then I will have more time to do something special with you later."

**To motivate cooperation the easiest reward
to give is more time with you.**

Whenever your child cooperates, the real consequence is more time later to do something they would really like to do with you. By reminding them of this simple truth, they will be quickly motivated to follow your direction. To make your reward even more effective, you may communicate it in ways that appeal more to your child.

REWARDS ACCORDING TO TEMPERAMENTS

Let's explore a few examples of communicating the same reward differently according to your child's temperament. With a more sensitive child, when describing the reward, focus on how it will feel. For example, "If you cooperate with me now, then I will have more time to do something special later. We could have a fun time picking flowers for Mommy in the garden. Mommy loves flowers. We could make a whole bouquet."

With an active child, when describing a reward, focus more on the details of action. "If you cooperate with me now, then I will have more time to do something special later. We could go play outside in the garden and pick a bunch of flowers for Mommy. We can even bring out the ladder and pick the blossoms from the tree."

With a responsive child, when describing the reward, focus more on the sensory details and tell a story. For example, "If you cooperate with me now, then I will have more time to do something special later. We could go out in the garden and pick the beautiful flowers for Mommy. We could make a bouquet with red, white, and yellow flowers. I bet we will even see some butterflies. When your mother sees her new flowers, she will light up with a big smile."

With a receptive child, when describing the reward, focus more on the timing. For example, say, "If you cooperate with me now, then I will have more time to do something special later. After school when we come home, we can pick flowers in the garden for Mommy. Right now, I need your help, and then later we will have time to pick flowers in the garden."

While framing the reward in different ways for your particular child will increase your child's motivation, just communicating the reward will still work. The simple message you convey is that time for me now means more time for you later. You help me now, and I will give you more later.

SAMPLE REWARDS

Here is a list of sample rewards. Take a few minutes to consider how you might communicate these rewards in a way that would work best for your children. Take into consideration their temperament. In addition, think of when you might need to use rewards and what rewards you think would work best with your children.

If you cooperate and pick up your toys now, then I will have time to play cards.

If you help me pick up your toys now, then I will have time to play a game with you.

If we clean up now, then we can do an art project.

If you pick out your school clothes tonight, then we'll have time to have waffles for breakfast.

If you get ready to go now, then we can come back again real soon.

If you get dressed now, then we can get treats right after school.

If you stop talking now, then we can walk the dog together.

If you get in the car now, then later I will play catch with you.

If you cooperate with me now, then I will do something special for you later.

If you do your homework now, then later we can have a little tea party.

If you eat your veggies, then we will have a dessert tonight.

If you come to dinner now, then we can sing songs after dinner.

If you come now, then you can play your game later.

When your children resist, instead of taking something away, give them a little more. Give them more support, so that they can once again feel their inner willingness to cooperate. Instead of using pain as a deterrent, use the possibility of more to encourage them.

ALWAYS HAVE SOMETHING UP YOUR SLEEVE

What makes giving rewards work is finding things that really motivate your children. Once you know what motivates your children, then always keep it up your sleeve. For one child, the only motivator you will ever need is "If you cooperate with me now, then I will have time to read you more stories." For another child, it might be, "If you cooperate with me, then we can make cookies in the kitchen together." Other children may need a variety of rewards. The secret of giving rewards is to pay attention to the things your children want most and use that to reward them.

The secret of giving rewards is to pay
attention to the things your children are
wanting most and use that to reward them.

If they really like stories, then hold back a little on reading stories. Certainly, don't stop altogether, but make sure that you don't overdo giving stories. Then reading stories becomes a more potent reward. Let's look at another example. When a child says, "Can we go to the park this week?" You say, "That's a great idea. If we have time we certainly will." At another time when they are resisting you can say, "If you cooperate now, then I will have more time and will take you to the park." Although you were already planning or hoping to take them to the park, you can now use that as a reward.

In a sense the very things you would take away to punish, can instead be given as rewards. If a parent would threaten not to take a child for a walk, then use going for a walk as a way to motivate the child. Instead of threatening with a statement like, "If you don't put these games away, then you can't play with them," say, "If you put these games away, then I will play a game with you later." The greatest and easiest reward to give is more of you.

The very things you would take away to punish can instead be given as rewards.

Rewards basically need to be somewhat logical, related, or reasonable. A logical reward is, "If you do this for me, then I will have time to do something for you." It is logical in the sense that if you do something for me, I will do something more for you. A related reward is, "It's time to go home for dinner. I understand you want to play and it's time to leave. If you come now, we'll come back again soon." The reward is related to the activity you want them to stop. A reasonable reward gives more according to the degree of resistance the child has. If the challenge is great for them, then offer them more.

Prepared parents always have a few rewards up their sleeve to pull out when their child is resisting. These are some common examples of rewards. Consider what rewards might be most appropriate to use.

A LIST OF REWARDS

We will have more time to do something special later.

You may ride your bike later.

We can pick flowers for the dinner table.

We can walk the dog together.

We can share a hot chocolate.

We can have a tea party.

We can play catch.

We can shoot some hoops.

We can bake cookies.

We can read three stories before bed.

We can go get a treat.

We can have dessert.

We can go swimming.

We can sing songs.

You can have a friend over.

We can go for a drive.

We can go shopping.

We can climb some trees.

We can go for a swing.

We can go play in the park.

We can do an art project.

We can draw.

We can paint.

We can go for a walk.

We can play cards.

We can cuddle.

We can watch a special video or show.

Giving a time warning can be a very good motivator as well. Particularly receptive children need more time to make transitions. A wise parent attempts to think ahead and prepare this child for changes. Instead of saying, "It's time to put on your jacket," say, "In five minutes we are going to leave for school. At that time, I want you to put on your jacket. If you cooperate and put on you jacket, then we will have a fun ride to school."

A wise parent attempts to think ahead and
prepare children for changes.

When lying with a child at night, if the child doesn't want you to leave, then you can say, "Okay in five minutes, I am going leave. If you cooperate and lie quietly, then I will stay the whole five minutes. If you keep talking, then I will have to leave now." Although having to leave now seems like a threat, it is okay because you have clearly given them a positive reward of your staying a whole five minutes longer if they lie quietly.

Before telling your kids to clean up their mess and come to dinner, let them know that in five minutes they will need to start cleaning up their mess to then come to dinner. Give them time to anticipate stopping, cleaning up their mess, and then coming to dinner. You could say, "You kids can play for five more minutes and then it will be time to clean up and come to dinner." In five minutes when you ask again they will be more cooperative.

The real magic of rewards is that, at times when nothing seems to work promising a reward will work. Without this clear insight and skill, positive parenting cannot succeed. Without the alternative of making a deal with your children through promising a reward, the only recourse is to threaten them with punishment.

RECURRING PATTERNS

When a child resists cooperating repeatedly, then offering the reward in advance is a useful approach. Once on a long plane ride, I had a difficult time getting my daughter Lauren to cooperate. From that time on, we solved the problem by preparing her for the trip. She loved a particular treat, so we promised her that treat if she cooperated the whole trip. By cooperating on the way to the plane and up to take off, she got a quarter of her favorite candy bar. Halfway through, she got the next quarter. Upon landing, she got the third quarter, and on arrival at our destination, she got the last quarter.

This plan worked beautifully each time. Before the trip

we would show her the whole candy bar. Her eyes would light up as we explained how much she would get at every stage of the trip. Although she was busy playing with things, never once did she forget to get her section of the bar. It was always in the back of her mind, keeping her focused on being cooperative. Besides giving her a reward, we also had the wisdom of providing her with activities to do during the trip. It is very unrealistic to expect a child just to sit and be happy doing nothing for a five-hour flight.

Besides being somewhat logical or related, a reward needs to be reasonable. If you are asking a child to do something you know they don't like doing, then it is reasonable to give a bigger reward. For example, if your child doesn't like certain guests that you like, then work out a deal such as this: "I know you don't like these people, but they are my friends. If you keep a friendly and polite attitude, I will do something really special for you. I will take you to the zoo next weekend." In this example, you give a big reward, because you are asking your child to do something out of the usual routine of life, and you know it is difficult for your child to do.

Children are more cooperative if we recognize what is difficult for them and give a little more because they cooperated. Whenever a recurring pattern of resistance is discovered, the best solution is to prepare in advance for it next time with a big reward.

REWARDING TEENAGERS

Rewards need to be age appropriate. Teenagers no longer need personal time as much, but they have other new needs. They need money and help. Once a preteen or teen is making and spending money, then it also can be used to reward.

Ideally, it's good not to use it too much of the time, but when used sparingly, it makes a big difference.

If a teen is resisting spending time in a situation, you could simply offer to pay them double their allowance or what they get paid for a day's work. If extra money in not available, then a parent can offer to drive the teen somewhere or help with one of their chores.

Some parents have found it's helpful to give their children rewards for improving their grades. Certainly, all children don't need this motivation. Better grades may be rewarded with more money, or they may be rewarded with extra privileges. Although privileges should be given when a teen earns the trust to have greater freedom, better grades could be a way to earn that trust. For example, by making better grades, a teen proves that he or she is more responsible, and therefore trustworthy and capable of staying out later.

DEALING WITH A DEMANDING CHILD IN PUBLIC

If your child throws a tantrum in a public place, recognize that you don't have the time to give the child what he or she needs on the spot to be more cooperative. This is when having your child's favorite candy bar handy is very helpful. You may not be able to listen with empathy to their feelings, but you can give a reward. You can remedy the situation by quickly offering your children a reward for cooperating. If you don't have something up your sleeve or in your purse, rather than resist your child and make a big scene, recognize what he or she wants and if possible, give it to the child. Even though this is placating the child, if only done occasionally, it is fine. It should, however, be a warning sign that you need to be tougher at home and not placate so much.

> When children are uncooperative in public it
> is a warning sign to be tougher at home and
> not placate so much.

Next time prepare your children by letting him know that you know it is more difficult to be cooperative in a grocery store check-out line. Let him know that you don't like waiting in long lines either. Then make a deal, say something like this, "If you cooperate with Mommy at the grocery store, then we will have time to come home and have a bowl of your favorite cereal." At the store, buy a box of that cereal to reinforce the deal. Throughout the shopping, let your children know how good they are doing and that soon they will be home eating a bowl of their favorite cereal.

REWARDS ARE LIKE DESSERT

When you offer children a reward, you are helping them once again connect to the part of themselves that wants to help. A reward doesn't make them cooperate. Instead, it is another way to nurture a child's natural motivation. Getting rewards or perks in life is like dessert. If all you ate was dessert, then you wouldn't get the nutrients you need from the meal. One of the reasons we give desserts at the end of a meal is that desserts alone will satisfy your hunger, and you will not be motivated to eat the foods that are good for your health. In a similar way, by relying only on rewards, children will lose their appetite to cooperate.

> By relying only on rewards children will lose
> their appetite to cooperate.

When an adult works only for the rewards then something is missing. They work only to get what they want and forget their underlying desire is to be of service. They don't care about doing a really good job. They only do what is required to get by. This is unhealthy.

On the other hand, it is just as unhealthy to be of service and not think about any reward or payment when your family is at home and in need. Successful adults think of both themselves and others. They care about making a difference, and they make sure they get what they need and want as well. Appropriately rewarding our children prepares them to be successful adults.

Appropriately rewarding our children
prepares them to think of both themselves
and others.

It is important for our children to learn that life is a process of give and take. If you give, then you get. To get more, you give more. Each time you ask your children to give a little more and promise to give them a little more, they are learning important lessons about life. They are learning how to make deals and negotiate. They are learning that they deserve more when they give more, and they learn to put off their immediate want in favor of some greater want in the future.

LEARNING FROM NATURAL CONSEQUENCES

For many parents, there is an implied assumption that if children don't cooperate then they are bad. They believe that good children should just automatically cooperate. Positive parenting recognizes that when children don't cooperate they are not bad, they are just not getting what they need. At times of resistance, a parent is required either to give them what they need or use rewards to motivate them in that moment.

Some parents mistakenly believe that good children should just automatically cooperate.

Some parenting approaches mistakenly recommend avoiding resistance by letting children do what they want and learn from the natural consequences. For example, if a child resists putting on his coat, let him go out in the cold without his coat and get sick. Then he will learn his lesson. This thinking is not correct. It just teaches the child that he can't rely on his parents for guidance.

Just as I was writing this passage my wife came in to share a relevant example. Lauren (age thirteen) had left her school paper in the computer at home. She had been working hard to finish her project on time and was very proud. Bonnie was going to bring this paper to her so she wouldn't lose points for turning it in late.

Some parents would say she needed to be taught the consequence of forgetting. She would not get the joy of turning her paper in on time. She would then learn from her loss. This is just old fear-based thinking. Instead, why not learn from her gain? Why not learn that her parents care, and they will do what they can to help? If your spouse forgot something, you would want to help him or her. You would do what you could. Our children need this support just as much or even more. Learning that they will get support from their family is much more important than learning how it feels to lose points on a paper you worked hard to finish on time.

The school of natural consequences would say this was

an opportune time to teach her what happens when she forgets things so that she will remember better in the future. It is true that she will be afraid of forgetting in the future, but fear is not really the best motivator. One can remember to bring something without fear. Positive parenting doesn't require fear to motivate children to remember. The experience of success also motivates children to remember.

Positive parenting doesn't require fear to motivate children to remember.

When we are afraid of making mistakes, we make more mistakes. Most people have experienced that fear tends to attract the very thing we are afraid of. For example, when I wear a new tie, I will often get a spot on it the first time I wear it. On the other hand, I get more compliments the first time I wear it.

If I am thinking how nice my tie is, more people notice it and compliment me on it. If I am very nervous about spilling food on it, it happens inevitably. The fear of making mistakes not only creates unnecessary anxiety in our lives, but it also causes us to make more mistakes.

An awareness of positive consequences is a better motivator; fear is not necessary to teach a child an awareness of consequences. Leave natural consequences up to nature; don't play God. Instead, parents should do their best to support their children. If you can't give them a particular support, don't; but if you can, do.

Leave natural consequences up to nature; don't play God.

The difficult issue here is: Am I sacrificing too much to give to my children? When parents deprive themselves, then they are giving too much and it will tend to make children overly demanding.

Giving too much can be easily corrected. Your children will let you know when you are giving too much. They will become overly demanding or you will begin to resent their demands. At this point, you need to back up from giving so much. This kind of adjusting is normal and to be expected.

THE FEAR OF REWARDS

Sometimes parents are afraid that if they give rewards their children will lose their natural motivation to cooperate. They imagine giving rewards and then having a child who says "What's in it for me," every time you ask him or her to do something. They go on to imagine an unwilling child demanding more and more in return for cooperation. While this nightmare is unlikely, it may occur if the child is not getting his other needs met as well.

Whenever you ask children to cooperate, a healthy part of them does ask "What's in it for me," and as long as they are getting what they need, they don't demand more. Children cooperate because all children are born wanting to cooperate in order to get the love they need. When children are aware of a need and trust that they will get the support they need, then they are more than willing to be cooperative.

When children get what they need they don't
demand more rewards.

As long as children basically get what they need, then they feel their needs and don't get lost in their wants. This awareness of their need for parental support makes children more considerate and cooperative. They don't ask for more all the time. They don't just focus on "what's in it for me" and demand more. As a general rule, children only focus on wanting more when they are not feeling what they really need.

Children primarily need rewards to overcome their resistance when they are not getting what they need in the moment. Offering to give a reward simply promises them that they will get more and suddenly they return to their natural willingness to be cooperative. Making deals and promising rewards is not caving in and giving children whatever they may want. It is actually the opposite: Giving a reward is asking the child to cave in to your wish and in return get more later. It is one of the most powerful ways to teach a child delayed gratification.

Sometimes giving rewards is not enough to minimize resistance and increase cooperation. When giving rewards doesn't work, then it's time to assert your leadership as parent or boss. When parenting becomes a little too child-centered and focused on giving children whatever they want, then the parents must assert their leadership to regain control. In the next two chapters, we will explore how this is done.

7

New Skills for Asserting Leadership

The greatest power parents have is the power to guide their children. Children are born wanting to please and cooperate with their parents. Children are already hardwired to respect the boss. Recognizing and using this power allows a parent to give up outdated practices based on fear and guilt. Without an understanding of how to use this power, the child takes control. Unless parents use their power to guide, they will lose control.

Children want to please their parents but, at the same time, they have their own wants and needs. When given the opportunity to feel and express their own wants, while also getting a clear message of what their parents want, children will ultimately seek to cooperate and yield to their parents' will and wish.

When parents use guilt or fear as a way to motivate cooperation, they weaken their child's natural willingness to cooperate. In response to a parent's anger, frustration, and disappointment, children may become obedient, but they will lose a part of who they are. Not only is their natural development restricted, but later in life they often become people pleasers. They do not have a healthy sense of themselves and tend to give more than they get back.

LEARNING HOW TO COMMAND

Before using a command the first step is simply to ask and not to demand. If the child resists your request, the second step is to listen and nurture. If listening isn't enough, then in step three we offer a reward. If offering a reward doesn't work, then step four is to assert your leadership and command. When the first three steps of asking for cooperation do not work, parents need to command their children, just as a general commands the troops.

To command is to tell your child directly what you want him or her to do. A command sounds like this. In a firm but calm voice say, "I want you to put your clothes away," or "I want you to get ready for bed," or "I want you to stop talking in there and go to sleep."

Once you use your command voice, you must remain strong. Using emotions, reasons, explanations, arguments, blame, or threats weakens your natural authority. Getting upset or trying to convince a child to cooperate at this point is a sign you do not feel confident in your role as general, boss, or parent. If the child has already resisted steps one, two, and three, then the parent needs to establish a clear message of who is the boss. By asserting leadership, you once again establish yourself as the boss. The child needs a strong leader to yield to.

Once you use your command voice,
you must remain strong.

Many parents command without consistently using steps one, two, and three. It does not work. Certainly, a parent doesn't always have to use the earlier steps, but if commanding is used too often, without the previous steps to invite and motivate cooperation, it loses its effect. In the past, children would submit to a parent's command, but today children need to be heard as well.

The most powerful assertiveness technique is
to repeat your command with the confidence
that the child will soon yield.

Children learn from previous experience that when you command, you will not yield. All negotiation is over. If they continue to resist, to assert leadership you simply hear their resistance with less empathy and repeat your command. The most powerful assertiveness technique is simply to repeat your command with the confidence that the child will soon yield. Ultimately, as the child continues to resist and you persist without resorting to emotions or reasons, you will prevail.

DON'T USE EMOTIONS TO COMMAND

If parents succumb to yelling, getting angry, displaying frustration, or making threats of punishment, they automatically give away their power to command. Using your upset emotions turns your command into a demand, and your positive position is weakened. You may be able to break a child's will and create obedience, but you will not be strengthening his or her natural willingness to cooperate.

Your power of commanding increases if you don't get emotionally upset. Remembering this will help you to stay calm. Subconsciously, we get upset because some part of us thinks it will make us more powerful and intimidating. Animals puff themselves up in combat to intimidate. Because positive parenting is not fear based, intimidation is not helpful.

> A clear and firm command repeated over and
> over without the tone of emotional distress is
> most effective.

Even some parent-effectiveness training programs have encouraged parents to share their feelings as a way to motivate children. Sharing feelings, even when done very peacefully, brings the parent and child to the same level, which slowly undermines the parent's position as boss. Although the intent to validate feelings is good, it is better to assist children in expressing their feelings and not burden or manipulate children with our own.

When children are resistant to cooperating, it is not the time for parents to share or reveal their feelings. Instead, this is the time to listen to the child's feelings. It is counterproductive for parents to share their anger, frustration, or disappointment. When children resist, it is time to identify and acknowledge what they might be feeling and what they want. After giving a child an opportunity to be heard and a possible reward for cooperating, it is time for parents to use their power as boss and to command their child. Commanding is well received after the parent has first listened to their child's resistance or objections.

IT'S OKAY TO MAKE MISTAKES

Certainly there will be times when a parent loses control and gets upset while making commands. As with all the skills of positive parenting, you don't have to be perfect for the program to work. But it is important to try. When you make mistakes by forgetting to hold back your emotions or you just lose control, the solution is to apologize later. It is okay to make mistakes.

Children don't need perfect parents, but they
do need parents who do their best and take
responsibility for their mistakes.

A little apology later on makes a big difference. It might sound like this, "I apologize for yelling at you. You did not deserve to be yelled at. Yelling is not a good way to communicate. I made a mistake."

Another way to apologize is by saying, "I apologize for getting so upset with you. I needed your cooperation, but didn't mean to get so upset with you. I was so upset because other things were bothering me as well. It wasn't your fault that I got so upset."

WHEN EMOTIONS ARE NOT HELPFUL

Whenever parents express negative emotions, children will feel that they have not measured up to a parent's expectations or are in some way inadequate or not good enough. They will feel they have failed in pleasing their parents. This feeling of failure or inadequacy will eventually numb children's willingness to respond freely. When a parent apologizes later for yelling, the child doesn't feel bad or unworthy. It is difficult to nurture the good in our children when we do things to make them feel bad.

Sharing negative emotions is helpful when we want to feel better. It is inappropriate to use our children to listen to our feelings to make ourselves feel better. It is healthy to need our children to cooperate, but not healthy to use them as a therapist or best friend. Children are still learning to deal with their own feelings; they cannot handle hearing their parents' feelings.

**An adult needs to go to another adult and
not to a child to get support for feelings.**

Whenever a parent expresses negative feelings, the children will eventually feel manipulated by feelings and stop listening. They will not only stop listening to their parents, but to their own feelings as well.

As with all other forms of manipulation, children tend to rebel later as teenagers to the extent they had to be obedient. Cooperative children don't need to rebel or disconnect from their parents to develop their independence. They can pull away to find themselves without giving up or rejecting the support of their parents.

YELLING DOESN'T WORK

Clearly, one of the worst forms of communication is yelling. The act of yelling implies that we are not being heard, so we are turning up the volume. To yell at children or teenagers gives the message "You are not listening." As a result, they just don't listen. Eventually, when you yell, they just turn off and hear nothing.

To yell a command is even worse. This means that they don't clearly hear what you want. Yelling disconnects children from their desire to be guided. Yelling is the weakest form of commanding, because it lessens your position as commander. Only when children clearly hear one simple message over and over do they give up their resistance to being led by the leader.

When you yell, you have stopped commanding and started demanding. It contains an implied threat: "You had better listen or else!" This kind of threat means you are demanding their obedience. Although making demands backed up with punishment has worked for centuries, its power cannot hold up in a free society. If you want your children to find the freedom in their lives to make their dreams come true, give them the freedom to cooperate. Don't demand it, command it.

MAKE YOUR COMMANDS POSITIVE

Although it is always best to command in clear and positive terms, often when you are ready to command the first thing that comes out of your mouth is a negative. If this happens, make sure to follow your negative demands or commands with a positive command. These are some examples of negative demands, negative commands, and then a corresponding positive command. Although negative commands are much better than negative orders, the positive commands are the best. Let's explore a few examples:

Negative Demand	Negative Command	Positive Command
Don't hit your sister.	I want you to stop hitting your sister.	I want you to be nice to your sister.
Stop talking.	I want you to stop talking.	I want you to be quiet now.
Stop fooling around and clean up your room.	I want you to stop fooling around and clean up your room.	I want you to clean up your room right now.
Don't talk that way.	I don't want you talking that way.	I want you to be respectful and say nice things.
Put your jacket on now.	I want you to stop fighting me.	I want you to cooperate with me and put your jacket on.

| You better listen to me or else. | I want you to stop playing cards and go brush your teeth. | Right now I want you to go brush your teeth. |

If you begin with a demand or a negative command from habit, then just follow your negative message with a positive message. For example you might say, "I *don't* want you to hit your brother. I *want* you to be nice instead." Once you have found your positive statement, don't deviate. If your child continues to resist, just repeat the positive statement. Let's look at some examples:

PARENT: Would you put you clothes away?

CHILD: Oh, I don't want to. I am too tired. I'll do it tomorrow.

PARENT: I understand you are tired and you want to do it tomorrow. I want you to do it now.

CHILD: But I am too tired.

PARENT: If you put your clothes away now, then there will time to read three stories.

CHILD: I don't care, I just want to go to sleep.

PARENT: I want you to do it now. Right now I want you to get up and put your clothes away. The discussion is over.

CHILD: You are mean.

PARENT: Right now I want you to put your clothes away.

CHILD: I hate you.

PARENT: Right now I want you to put your clothes away.

CHILD: (getting up to put the clothes away) I can't believe you are so mean.

As you see the child beginning to cooperate with your will and wish, simply leave him alone for a few minutes or quietly watch. Then come back in and thank him in a friendly way, making it clear from the tone of your voice that you don't resent occasionally having to push so hard. Even if you have to go to all this trouble, children should not feel that they don't get credit for eventually responding to your will.

Many parents will not appreciate a child's compliance when they have had to keep commanding. This is in some ways similar to the woman who doesn't give her husband credit for doing something nice if she had to ask. Parents need to always remember that children do their best, and whenever they move in the right direction we should acknowledge and appreciate it.

In this example, the parent could say then or the next morning, depending upon when they feel it, "I know you were very tired. I appreciate you for cooperating." When you don't hold their resistance against them, they won't continue to resist you by resenting your commands.

When you don't hold it against them, they won't hold it against you.

Some parents are afraid that their children will not love them if they command their children in this way. There is nothing further from the truth. Children need a strong but loving parent. They need a parent to motivate them in this way. They will grumble at you, but they will come back very quickly to loving you. One of the five messages of positive parenting is that it is okay to say no, but always remember mom and dad are the bosses. Assertive leadership through commanding puts the parent back in charge.

In the example above, there is basically a battle of wills. If you persist in repeating your command and avoid getting into a big discussion or debate, you will win. Once you have won a few battles of will, your child will be more cooperative. Nice parents are usually afraid to seem so mean, but this is necessary. You are not being mean in an unloving way. You are demonstrating that you mean what you say.

COMMAND BUT DON'T EXPLAIN

Besides using too much emotion, the other common mistake parents make is to justify their commands with explanations. If the child asks in a nonargumentative tone, it is fine to explain why you want something done, but if she asks in a challenging way, let her know you will be happy to talk about it later. If there is not much time just say, "We can talk about it later, but right now I want you to stop hitting your brother. I want you both to cooperate and get along."

Giving reasons is a way of giving up your command. If children could understand the difference between right and wrong, then they wouldn't need you. If they could comprehend what was good or bad, then you wouldn't be needed to direct them. When you are on an equal level with someone who is reasonable, reasons work. Children do not develop the capacity to reason until about the age of nine, and they are not on an equal level until they are ready to leave home sometime around the age of eighteen.

Giving reasons is a way of giving up
your command.

Children have the latent ability to know what is right or wrong, but it becomes awakened by cooperating with your requests and not by listening to your lectures. When you ask a little boy to stop hitting his brother and then command him with your "I want" statements, he will respond. When he sees the smile on your face because he did what you asked, then he begins to learn what is good and what is right behavior.

Children learn right and wrong by
cooperating with your requests and not
by listening to your lectures.

Once you begin commanding, avoid stating the rules or giving reasons to back you up. Once a parent begins to repeat a command, the time for negotiation is over. Children have the right to question and negotiate during the first three steps, but once you begin to command, negotiations are over. It weakens your power if you digress to have a discussion about why the child should do what you want. At this point, the best technique is repetition of the command. The child still has the right to resist, but you are still the boss. As you persist in commanding, the child will automatically begin to yield to your point of view.

When you state a command, the only reason children should cooperate with your request is because you want them to. If you didn't give them some opportunity to resist, then you are simply demanding that they be obedient instead of cooperative. Deep inside every child the strongest motivating force is the desire to please and cooperate with parents or the primary caretaker.

Children are born ready to follow our guidance; we just have to give them a chance. The desire to please and cooperate is the prime directive. By directly expressing our will, we ultimately awaken their will. Eventually, by expressing their resistance to your will, their deepest will to follow your guidance emerges and they yield.

By first asking and then hearing any resistance before they yield to your point of view, they will not be surrendering their will in blind and mindless obedience but, instead, will be adjusting their will to their own prime directive, which is to follow your guidance and make you happy.

Children don't need reasons. Instead, they need strong leadership. They need to be reminded that *they* are not in control, but that you are the boss. Trying to explain what is right or wrong, good or bad when they are resisting you only weakens your power to command. Even with teenagers who are capable of reason and abstract thought, when it comes down to a command, the reason they should cooperate is that you are the parent and you want them to.

COMMANDING TEENAGERS

I remember when I first experienced the power of commanding. Even before I raised my children, I was teaching a workshop for children of broken families. Many of these children were very unruly and uncooperative, which is why their parents enrolled them in my workshop.

At a certain point, the oldest child in the group, who was about fourteen, became very resistant to my requests. I proceeded to send him into the next room for a time out. He resisted and said, "What are you going to do about it?"

Although back then I didn't know different positive-parenting skills, I realized that the threat of punishment was not going to work. As he glared at me, I could see that whatever I said would be countered with a "so what."

Since he was already being punished to the maximum in his life, another punishment meant nothing to him. He had stopped caring and was defiant. He was also much bigger than I was. I didn't really know what else to do, so I just looked him in the eyes and continued to command him in a clear and firm voice saying, "I want you to take a time out for about fifteen minutes." Our conversation went on as follows:

TEENAGER: And what if I don't?
ME: I want you to go in the next room and take a time out for fifteen minutes.
TEENAGER: You can't make me.
ME: I want you to go in the next room and take a time out for fifteen minutes.
TEENAGER: You are a wimp, you can't make me.
ME: I want you to go in the next room and take a time out for fifteen minutes.
TEENAGER: What are you going to do if I don't?
ME: I want you to go in the next room and take a time out for fifteen minutes.

He made an expression of disgust and walked away into the other room.

After about fifteen minutes, I went into the other room and, in a friendly manner, said, "If you want to join us, you are welcome, but if you need more time alone, I will certainly understand."

He silently nodded his head as if he would think about it. I left the room in a friendly way, and a few minutes later he came out and joined the group. This experience was great preparation for successfully dealing with the inevitable resistance I encountered with my own children.

You can see that if I had reacted to his comments or answered any of the questions, my position would have weakened. Ultimately, every child, until ready to leave home, needs parents to assert leadership and be the boss. Ultimately, when faced with a caring adult whose commands are clear, an unruly child will eventually yield without threats or disapproval.

REASONS AND RESISTANCE

When children continue to resist going to bed, giving reasons doesn't help. Telling them it is late or they need their sleep is not going to win them over, nor will it teach them anything. If children ask "why" and they are not being argumentative, then explanations are useful, but they are counterproductive at times when children are resisting. At a certain point, the only good reason they should do something is because you are the parent. I remember getting a funny T-shirt for my wife with this message. It said, "Because I am the parent, that's why."

"Because I am the parent, that's why"
is the best response to children's challenge
to commands.

Teenagers will often drive their parents to the limits of frustration. They will question and challenge whatever a parent commands. Parents try to be reasonable and lose. Children and teenagers just keep asking why, yet each time you answer, you become weaker. With each answer, you get further and further away from your true power. All they have to do is keep asking, which is what they do.

Let's explore an example. Carol wants to watch TV, and her mother wants her to do homework.

MOTHER: Carol, I want you to turn off the TV.

CAROL: Why?

MOTHER: You need more time to do your homework.

CAROL: I don't have any homework today.

MOTHER: But you have projects due and you wait to the last minute and then complain that you have too much to do. If you don't have homework, then this is a good time to get ahead on your science project.

CAROL: I've done everything I can. I can't do anymore until I get the pictures developed and that will be several more days.

MOTHER: Well, you have been watching too much TV.

CAROL: No, I haven't.

MOTHER: Yes, you have. You have been sitting here all afternoon.

CAROL: You haven't even been here to know. You just got back.

MOTHER: Yes, and you were watching TV before I left.

CAROL: But I wasn't watching TV all this time.

MOTHER: Sitting in front of the TV too much is not good for you. You should get outside. It is a beautiful day.

CAROL: I don't want to go outside. My legs are sore from gym today.

MOTHER: You had better listen to me, young lady. You are going to lose all your TV privileges if you don't watch out.

CAROL: You are so mean and unfair.

MOTHER: You are this close to being grounded with no TV.

CAROL: I don't care.

MOTHER: Okay then, you are grounded for two weeks with no TV.

These kinds of arguments and fights can be avoided if parents don't get off the track trying to convince children or teenagers of the merits of what they are saying. If children or teenagers are receptive to your ideas, it is different, but, if they are resistant to your request, they will continue to resist your ideas even though they may be good.

Here is an example of what a mother using positive-parenting skills could do to avoid fighting or arguing with her teenager or child.

A BETTER WAY OF COMMANDING

Step 1: Ask (Don't order)

MOTHER: Carol, would you turn off the TV?

CAROL: Why? This is a great movie.

MOTHER: What is it?

CAROL: It's Sherlock Holmes.

MOTHER: That is a great movie (pause), but I want you to turn off the TV. You've been watching a lot of TV lately, and I want you to do something else.

CAROL: Like what?

MOTHER: You could work on your homework or do something outside.

CAROL: I don't want to. I just want to watch my show, and you are disturbing me.

Step 2: Listen and Nurture (Don't Lecture)

MOTHER: I understand that you just want to watch your show and you don't want to do homework or go outside (pause), but I want you to turn off the TV and find something else to do.

CAROL: I don't want to.

MOTHER: I know it is disappointing, but now it is time to do something else.

CAROL: But I will miss the rest of my show.

MOTHER: I'm sure it will be on again soon.

CAROL: No, it won't.

Step 3: Reward (Don't Punish)

MOTHER: If you turn off the TV now, then I will take you to get a video tomorrow.

CAROL: I don't care about renting a video. I just want to watch my show.

Step 4: Command (Don't Explain or Get Upset)

MOTHER: I want you to turn off the TV now.

CAROL: But I don't have anything else to do.

MOTHER: I want you to turn off the TV now.

Carol gets up and turns off the TV. She storms out of the room. About fifteen minutes later, she comes back, acting like nothing happened and asks to play cards. Her mother happily agrees. No mention is made of the little battle. All is forgotten and forgiven.

INCREASING COOPERATION

The result of using positive-parenting skills is that your children will be more cooperative in the future, instead of—as the result of arguing, fighting, and punishing—increasing resistance, which gradually turns into unhealthy resentment, rejection and rebellion. When you command your children using emotion, logic, reasons, argument, or threat, it only weakens your position and strengthens their resistance in the long run.

When you use these first four steps of positive parenting gradually over time, you only need to ask and your children will cooperate most of the time. Inevitably, those times will surface again and again when you may need all four skills. As you use these four skills, they become easier. Not only do they work to create cooperation, but they become clear and definite ways to nurture your children to be their personal best.

If it seems like a lot of work, that is just because it is new. Learning any new skill can be a bit overwhelming, but with a little practice it becomes easy and automatic. Raising children is always challenging, regardless of the child. Positive parenting is not any more difficult. In the long run, it is much more fulfilling and effective as well.

Being a parent is filled with waves of challenge. We can either surf those waves or get knocked around again and again. Commanding our children may seem a little tough for some soft-love parents, but it is the clear and positive alternative to demanding our children into obedience with threats and shame.

CHOOSING YOUR BATTLES

Before commanding take into consideration your child's temperament.

Sensitive children also require greater assistance. Rather than expecting a sensitive child to clean his room, ask him to join in to help you. In this way, by doing things together they will slowly become more independent. After commanding them to help clean their room just join in and start cleaning.

Responsive children may feel cleaning the whole room is too much and then feel the need to move on to something else which is easier and less time consuming. Parents need to give these children an opportunity to move from one thing to another. Remember they are like butterflies and need to keep moving around.

Before commanding a responsive child, parents must first attempt redirecting a child. When they are unwilling to clean their room, try directing them just to clean up one thing and then go on from there. Sometimes it is enough if they do a few things, and you do more. Eventually, they will want to do more, but it takes time.

Receptive children generally don't require commanding. They tend to be more accommodating. If they resist, it is often because the parent expects them to make a change before they are sufficiently prepared. Once they receive the reassurance and preparation they need, they will be more willing to accommodate.

Rather than commanding receptive children, parents need to be more understanding of the child's need for rhythm and repetition. These children don't respond well to sudden changes, interruptions, or demands.

Active children respond best to commanding in private. Take them to the side or into another room to command their cooperation. They pride themselves on doing the right thing, on being in control, and may become unnecessarily defensive or resistant in front of others.

8

New Skills
for Maintaining
Control

When a child defies or rejects parental control, instead of recognizing this behavior as bad or wrong, positive parenting simply acknowledges that the defiant child is out of control—out of his parents' control. Instead of judging, punishing, or lecturing the child, all that is required is to bring the child back into control. When children are out of control, a parent is required to contain or restrain them from continuing to defy or reject the parents' control.

The purpose of a time out is not to threaten or punish a child in any way. It is simply a way to help children experience that they are once again in their parents' control and that they prefer it that way. Children need to push up against limits in order to find acceptance and begin to cooperate again.

God made children little so that when they
go out of control we can pick them up and
put them in a time out.

When children misbehave, they have often just forgotten that their parents are the boss and that they actually want it that way. When children are out of their parents' control, they have disconnected from their natural ability and willingness to cooperate. When children are not getting what they need to feel their desire to cooperate, they eventually become disconnected from their parents and go out of control. Children need guidance. When they stop feeling their need to be guided, they spin out of control.

Inviting cooperation, listening to and nurturing a child's needs, and giving rewards keeps a child connected to his or her willingness to cooperate. When stress increases for the child or the parent, this inner connection is temporarily broken. Like a car out of control, the child will inevitably crash.

Under stress children go out of control, like a
speeding car without a driver.

When parents lose touch with their willingness to cooperate and begin demanding obedience, children automatically follow suit and lose touch with their willingness to cooperate. A stressed parent feeling out of control can easily push a child out of control as well. Conversely, a stressed child feeling out of control can easily push a parent out of control, unless that parent clearly knows how to regain and maintain control.

THE NEED FOR TIME OUT

With these new tools for creating cooperation, parents can maintain control within themselves, which in turns helps children stay in their control. Inevitably, some kids will go out of control on a regular basis, and the positive parent is prepared to deal with it. Almost all children need to take time outs on a regular basis to learn how to regain control when emotions become too strong to control.

**Time outs are needed to regain control when
emotions become too strong.**

Even many mature adults don't know how to handle their inner emotions at stressful times. We cannot expect children to do so. By teaching thousands of adults to manage their inner feelings, I discovered many of these skills for creating cooperation. When a parent is stuck feeling resentful, anxious, depressed, indifferent, judgmental, confused, or guilty, the answer is always to look within to manage the negative emotions.

One of the major reasons there is so much more domestic violence in the West today is the lack of emotional control. In a free society, it is inevitable that feelings become richer when supported and more volatile when not supported. The first skill for resolving conflict in a relationship and ending all violence is recognizing that when feelings become strong, defiant, or rejecting, it is time to take a time out and cool off.

**In a free society, it is inevitable that feelings
become richer when supported and more
volatile when not supported.**

Adults who lose control and act out violently do so primarily because they never learned to take a time out to feel and release their negative emotions. This basic ability is needed by children, teens, and adults as well. The difference between a child and an adult is that a wise adult should know when she needs to take a time out and a child doesn't. Children at nine or ten years old, who have been raised with time outs, will automatically begin taking time outs on their own whenever they become stressed, negative, or argumentative. It is not a difficult skill, but it takes practice.

Sometimes the reason a time out works is that it puts the parent back into feeling in control. When a parent begins to feel out of control, it pushes a child out of control. By giving a time out, the parent has a chance to cool off and once again feel in control. This in itself is often what the child needs. A frustrated, demanding parent can easily spin a child out of control. The option of giving a time out not only puts the child in control but helps the parent as well.

A frustrated, demanding parent can easily
spin a child out of control.

When a child is not willing to cooperate with a command, then it is time for a time out. At this point, a time out is an opportunity for the child to blow off steam and throw a tantrum. Children need to feel their resistance to life's inevitable limits and boundaries. They need to push up against those limits and feel their resistance. This pushing or resistance helps a child develop a strong sense of self. Ultimately, all the positive characteristics of the true self emerge. First, the negative emotions under the resistance must be felt and released.

Time to push against the limits is needed to help children experience the different layers of feelings underneath their resistance, resentment, rejection, or rebellion. Resistance is released when we are able to feel and release the three underlying negative emotions of anger, sadness, and fear. In a similar manner, adults can release the negative blocks of resentment, guilt, self-pity, when they take a time out to explore, feel, and release their negative emotions.

HOW NEGATIVE FEELINGS GET RELEASED

When a child is stuck in resistance and will not respond to the four steps of creating cooperation (see Chapter 7), then step five is required. Being contained in a room actually assists children in becoming aware of the deeper levels of their emotions. By feeling the three deeper levels of anger, sadness, and fear while the child is being cared for, the negative feelings are automatically released.

Giving a time out allows a child first to feel anger and frustration. Then, after a short period of time, the child will begin to cry and feel sadness or hurt. A little later, the child will feel his or her underlying fears and vulnerability. Within a few short minutes, all this drama will lift away and suddenly once again the child will be miraculously back in your control.

Within a few short minutes of time out, all
the emotional drama suddenly lifts away.

In a time out, by first feeling the primary emotions that come up when contained, children return to feeling their needs. Prior to a time out, children are acting out of control because they have forgotten their need for loving guidance and their desire to cooperate.

By putting children in a time out, they get a chance to

resist doing what you want. Then a switch turns and they begin to feel their emotions. Instead of just being emotional, they "feel" their emotions. The act of resisting the time out actually increases their ability to feel. With this increased awareness of what they feel, they begin to experience their need for parental love, understanding, support, and guidance. As they feel their need for love, their desire to cooperate is activated once again.

Children are from heaven. They are born wanting to please their parents in order to get what they need. Children need love and support to survive, so they are born with a willingness to cooperate and please in order to get that love. This healthy desire is activated whenever children reconnect with their feelings. Increasing children's awareness of their feelings draws out their need for love and support, which then awakens their desire to please and cooperate.

Sometimes these levels of feelings come up when a child resists our request for cooperation. At other times, feelings come up in tender conversation. At other times, when children are not getting what they need or when life is just too stressful, they will require more time outs to feel and move through the three levels of emotional resistance: anger, sadness, and fear.

THE IDEAL TIME OUT

The ideal time out is accomplished when a parent puts a child in a room and holds the door shut. It is a natural expression of resistance for a child to try to get out. Remember, children are supposed to resist. Locking the door and leaving a child creates feelings of abandonment. Being present on the other side of the door, at least in your children's early experiences with time outs, is very important for some children. After many time outs, a child will not try to get out.

The time needed is generally one minute for each year. A four year old goes in for four minutes; while a six year old goes in for six minutes. When parents first hear this, they can't believe it will work for their child, but it does. It works for all children and all ages starting at two years old.

After fourteen, giving time outs is rarely necessary. If you didn't raise your children with many years of time outs, your teenagers will still require them, particularly at those times when they are being disrespectful or will not listen to your commands.

Let's explore what happens in an ideal time out for a four year old. At first, the four year old resists, and you may even have to carry him into the room. It can be his room or any room. At first, he will get angry, throw a tantrum, and try to get out. After about two minutes, he will stop trying to get out, but will surrender more to the limits and begin to cry. After another minute, he will shift to the more tender and vulnerable feelings of fear. At this point, he may even put his little fingers under the door and beg to come out saying, "Please, please let me out."

At this point, it is fine to assure him that he only has one minute left and soon will be out. It is actually fine to assure children at any time during a time out. You might let them know repeatedly that you are not going anywhere, that you are just on the other side of the door, and that soon they will be able to come out.

EXPLAINING TIME OUTS

When children ask why they have to take a time out, the simple answer is this, "When we go out of control, we need a time out." It is neither accurate nor helpful to say that a child needs some time to think about what he or she did wrong. Thinking in a time out is not necessary. All that is needed is to feel the emotions that come up and automatically the child comes back into control.

Children don't need to think about what they did wrong. When parents focus too much on right and wrong with children, the only thing children learn is to feel guilty. Instead of telling your children what is right or wrong, a better approach is simply to ask your children for specific behaviors and increased cooperation. As children cooperate, they automatically learn what is right or wrong, good or bad; they don't ever need to be told they are bad or wrong.

As children cooperate, they automatically
learn what is right or wrong, good or bad.

Giving time outs replaces the need to punish or spank children. A time out connects children to their feelings and their need to cooperate, but in a completely different manner. When children are punished or spanked, they punish themselves or others when they go out of control. Children who take regular time outs do not punish themselves or others in order to regain control.

Giving time outs replaces the outdated need
to punish or spank children.

Adults who were punished continue to punish themselves or others. Adults who were not greatly punished have a higher sense of themselves and their worth and are much more successful in getting what they need while giving to others.

With regular time outs, children learn to manage their inner feelings. When life's inevitable winds push them out of balance, they automatically take a time out, let go of their negative feelings, and return to their true self. They become more loving, happy, peaceful, and confident, and they are naturally motivated to cooperate rather than demand, submit, or manipulate.

FOUR COMMON MISTAKES

Many parents think they are giving effective time outs and complain that it doesn't work. Time outs work, but they must be used correctly. These are the four most common mistakes parents make in using time outs:

1. They use only time out.

2. They don't use time out enough.

3. They expect their children to sit quietly.

4. They use time out as a deterrent or punishment.

By setting the right parameters regarding time out, your children will come back into your control. They will once again reconnect with their inner prime directive, which is to follow your guidance and cooperate. Let's explore each of these four mistakes in greater detail.

1. Too Much Time Out

Just giving time out and not applying the other skills of positive parenting will eventually lessen the effectiveness of taking a time out. Time outs are to be used as a last resort or at times when you just don't have time to move through the other four steps of positive parenting. To be cooperative and flourish, children have other needs besides their need to push up against the limits of a time out.

**Time outs are to be used
as a last resort.**

Even though the body needs vitamin C, it has other needs as well. Vitamin C alone will not keep a body healthy. If you are vitamin C deficient, then it will make a big difference to your health because your body needs it. If you have enough of the other required vitamins, you will notice an immediate improvement in health. If you only get foods with vitamin C and ignore your other needs, then even vitamin C will not make much difference in keeping you healthy. In a similar way, each of the five steps of positive parenting is equally important to create cooperation.

2. Not Enough Time Out

While some parents rely too much on time out, others don't use time out enough. They complain their children just will not listen. For example, one mother complained, "I ask him to stop jumping on the bed and he just laughs at me and keeps jumping."

This is a clear sign that this mother is not giving enough time outs. A time out gives a parent control. If a child just laughs and ignores you, then clearly this child is out of control and needs more time outs. This parent needs to pick the child up and move him into another room for a time out.

If a child just laughs and ignores you, then
clearly this child is out of control.

Some parents conclude that a time out doesn't work, because the next day the child goes out of control again. These parents mistakenly assume that if time outs were working, their child would always cooperate and not resist. A time out does not break a child's will and create obedience. It strengthens a child's will, but also nurtures her willingness to cooperate.

Children will be children and will go out of control. The regular need for time out doesn't mean it doesn't work. Active children in particular and little boys will tend to need more time outs. If your child needs more time outs, it doesn't mean something is wrong with your child or your parenting approach. It is simply what your child needs at this stage in their development. There is no right number of time outs. It could be two a day, two a week, or two a month, or two a year. Every child is unique.

If your child needs more time outs, it doesn't
mean something is wrong with your child or
your parenting approach.

Soft-love parents use time outs, but generally not often enough. Instead of commanding cooperation, soft-love parents tend to cave in and give too much to the child. They can't bear to hear the child scream, so they routinely placate the child.

Their child may so fiercely oppose a time out that parents will do anything to avoid a confrontation, even if that means giving in and doing what the child wants. When children become too demanding or bossy, it is a clear sign that the parents are not maintaining control with an adequate number of time outs.

3. Expecting Your Child to Sit Quietly

Some parents misunderstand the whole purpose of a time out. They expect the child to sit quietly and cool off. Instead of giving a time out for the child to feel and release negative emotions, these parents discourage the child from getting upset. They give messages like, "If you continue to resist then your time out doesn't begin until you stop."

A time out works because it gives children the opportunity to resist more. Encouraging your children to give up resisting and sit quietly instead is not a time out. Children should feel free to resist a time out. They are not supposed to like it, and they are not supposed to be quiet.

Children should feel free to resist a time out.

There is nothing wrong with giving a child a cooling off period. This is a form of redirection and one of the nurturing skills. If children are getting hyper and resist cooperating, then having them cool off by sitting in a corner or on a special bench is fine. This is similar to having children take naps when they become too fussy and resistant to direction.

A cooling-off period is not the same as a time out. In a cooling-off period, children are encouraged to be quiet and may even be rewarded for taking some time to cool off. Cooling off does not encourage children to move through their feelings. The first step in learning to manage negative feelings is to feel them and release them. As children get older (around the age of nine), they are able to feel and release emotions without a time out.

Just cooling off does not encourage children to move through their feelings.

A parent might say to an argumentative teenager, "This isn't working. I want you to take some time in your room to cool off and then we can talk again." This cooling off is all the teenager may need. While this is similar to a time out, it is still different; it is just directing your teenager to another activity to lessen resistance.

If the teenager resisted and the parent commanded and the teenager still resisted, then it would be a time out. The teenager would eventually storm away to his room. At such times, the parent must be very careful not to reprimand the child for resistance, but to continue commanding until the child goes into his room. When the teenager comes out, he will seem to be a different person.

4. Using Time Out as Punishment

The fourth mistake parents make is to use time out as a punishment. Although children may feel they are being punished by time outs, a parent must be careful not to use a time out as a punishment. As we have already discussed, fear-based parenting uses the threat of punishment to deter children from misbehaving. The threat of a time out can easily be misused to control a child. Quite often, parents will give a warning or signal saying, "If you don't stop you will have to go take a time out." This warning will stop children in their tracks as if the parent had said, "If you don't stop, I will tell your father when he gets home" or "If you don't stop, I will spank you."

Threats have worked for centuries, but in a free society fear-based parenting comes back to haunt us. The more parents use punishment, the more their children will rebel later. Many adults today still can't connect or don't feel a desire to connect with their parents because they were punished.

HUGGING DAD

When I became an adult, I had a great relationship with my dad. When I started teaching seminars on relationships, he was the first member of my family to attend. He would fly out to California from Texas to take my seminars. In these seminars, one of the exercises was learning to hug people.

I noticed that when I would hug people I could easily feel a warm connection. Yet, when I hugged my dad, although we had a loving and supportive adult relationship, I could only feel a slight connection. It was as if there was a wall separating us. I could hug a stranger and feel a lot more warmth and connection.

I asked my friends, who had hugged my dad, what their experience was. They said he was warm and friendly. They felt the connection, but I didn't. I realized that this was from years of disconnecting with my natural desire to be guided by him. I was a good, obedient child, because of the fear of punishment.

As an adult, it took another ten years of self-therapy and participating in workshops to heal unresolved feelings before I could feel my connection with my dad when I gave him a hug. If he had known the techniques of positive parenting, he would have been happy to use them, and I would not have needed to heal these unresolved issues.

Using a time out as a threat will work in the short run, but in the long run the delicate and tender desire to cooperate with the parent is gradually weakened. Certainly, a time out is a much better punishment than taking things away or hitting, but it is still less than the best for your child.

ADJUSTING YOUR WILL VERSUS CAVING IN

Adjusting your will to give children what they want is not a great crime. It gives a clear and healthy message that parents listen and learn, and it demonstrates healthy respect and flexibility. Adjusting your will becomes caving in when your motive is to avoid confrontation. It is not healthy to cave in to your child's demands.

Caving in will spoil children and make them more demanding. Remember, it is not giving children more that spoils them; it is giving them more to avoid confrontation. Children need to experience again and again that the parent is the boss. Children become spoiled when parents let them be boss. Spoiled children have disconnected from feeling their need for the parent to be the boss.

> It is not giving children more that spoils them; it is giving them more to avoid confrontation.

When children don't get enough time outs, they become prone to more intense tantrums. This means that when the child finally gets a time out, he or she will throw an even bigger tantrum. Eventually, with regular time outs, the child will come back into balance and be more cooperative rather than demanding. If you have caved in and spoiled your children, they can become unspoiled again by giving more time outs. Children are never really spoiled, just out of control.

After a while, you will easily sense when your child just needs a good cry. Crying is one of the most effective ways to release stress and feel better again. When you experience a great loss, grieving is an essential part of finding acceptance again. When our children experience their disappointments and losses, though these traumas may seem small to us, they are big to children. Children also have a need to cry or grieve as a way to accept the limits and boundaries of the world.

Sometimes a child just needs a good cry
to feel better.

Some parents mistakenly assume that they are hurting their children because their children will cry. Without the insights of positive parenting, they mistakenly conclude that it is too cruel a measure. This same parent will often turn around and spank, yell, or punish a child when nothing else works.

A time out for a few short minutes does not hurt a child, but it does help to bring up the accumulated feelings that are right under the surface. A time out brings up the painful emotions that need to be felt in order to be released. Although children never like taking time outs, they need to have good cry and come back into balance.

WHEN TO GIVE TIME OUT

Whenever you threaten to give a time out, you are using it as threat of punishment. Instead of threatening to give a time out, a parent should just give them. The best time for giving a time out is after a child has had the opportunity to respond to your commands. If, after you have repeated a command a few times, the child continues to resist, then he or she needs a time out. The time out is not given as a punishment, but because the child needs it. Even though the child may consider the time out a punishment, as long as you don't use time out as a punishment, it isn't a punishment.

If you warn children that if they continue to resist they will need a time out, you are still using time out to threaten children into obedience rather than using your rewards and commands to motivate cooperation. This fear-based approach only weakens your ability to command your children in the future.

THREE STRIKES AND YOU ARE OUT

After the age of nine, this circumstance changes. Children are now more capable of containing their feelings and don't need time outs as much. They have learned to feel their emotions and let go of them. In this case, the child is given the opportunity to find within herself the ability to come back into control. Children get three strikes, and then take a time out.

When a child is resisting and not responding to a command, the parent can simply say, "That's strike one." Strike one is a code that means that if she can pull herself together on her own and cooperate again, then she doesn't need a time out. If within a few minutes she continues to resist, the parent says, "Strike two." This means the child has one more chance to cooperate. If within a few more minutes the child continues to resist, then the parents say, "Strike three," and the child takes a time out.

After explaining this to a nine year old, you can create

your own code. You may wish to pull on your earlobe instead of saying "strike one" or you may just simply hold up one finger, two fingers, and then three fingers. Once you begin to use this approach, you need to be consistent. Before getting a time out, your children will expect to get their signals, which is good. Learning to pull ourselves together without having to suppress our feelings is an important skill. For children to pull it together, they need a few tries or strikes.

WHEN TIME OUT DOESN'T WORK

Some parents complain at workshops, "Time outs don't work for my teenager. I can't even get him to do it. He just laughs and walks off. I have to take something away to punish him." This is just another example of the defiance teenagers today feel in response to all the punishment of the past. Fortunately, after years of using fear to motivate children, you can begin using the five steps of positive parenting and the five messages. They will begin to work immediately.

In this example, the parent needs to begin with the first four skills instead of just using time outs as another way to punish. Giving time outs work—when used together with the other four skills. Within a short period of time, your teenager will be much more cooperative. Time outs are most effective for young children. Learning to listen without giving advice and using rewards to motivate are much more effective skills to use with teenagers.

Time outs work by creating an opportunity for a child to resist. Some children are so disconnected from their inner desire to please that at first they don't resist a time out. They are happy to take a time out and would prefer being alone to being with their parents. These children have disconnected from their desire to please or be guided. Some children just give up trying to please their parents or feel so controlled and manipulated that they don't want their guidance either. They don't mind being alone and like showing their defiance in this way. Often, when children resent their parents, they happily go to their rooms to show they don't care.

In this example, parents need to look at steps one, two, and three (see list below) to nurture their child back to feeling a desire to please and cooperate. Then this child would respond differently to time outs and benefit more from them. Even though the child appears to be happy, he or she is still being put in a situation beyond his or her control and is thus back in your control. If a child likes going to his or her room, use another room where he or she can't play with their games, toys, stereo, or talk on the phone.

It would be fine to send children to their room to play a game, but that would not be a time out. If they resist taking time outs, then it is fine for them to use their games or toys during their time out. If they are happy about time outs, then put them in the bathroom or another room.

WHAT MAKES THE FIVE SKILLS WORK

The five skills of positive parenting work today because the world is a different place and so are our children. In a free society, we must adapt our parenting approach. In summary these skills are:

1. To create cooperation, ask don't order.

2. To minimize resistance and improve communication, listen and nurture—don't fix.

3. To increase motivation, reward—don't punish.

4. To assert your leadership, command—don't demand.

5. To maintain control, give time outs—don't spank.

These five skills work to awaken our children's willingness to cooperate. The fuel that makes these skills work are the five positive messages (see "Introduction"). Without these skills, we cannot effectively put the five positive messages into action, but it is the five messages that make the skills work. The five skills and five messages are interdependent.

The first message—it is okay to be different—nourishes our children's need to feel loved and special. Without our understanding and accepting how each child is different, children cannot get the nurturing they need to be cooperative.

The second message—it is okay to make mistakes—is essential for children to feel good about themselves and continue to be motivated to please their parents in a healthy way. If mistakes are not accepted, then children either give up trying or give up themselves in the process of trying.

The third message—it is okay to have negative feelings—makes it safe for children to grow in an awareness of what they feel inside. This awareness is essential for keeping children in touch with their healthy need for parental guidance and approval, which in turn triggers their willingness to please and cooperate.

The fourth message—it is okay to want more—opens the doorway for children to develop a strong sense of self and direction by knowing what they want. Children who know what they want are most easily motivated by the possibility of more. They not only want more, but learn how to delay gratification when they can't get it right away. When children have permission to want more, they quickly respond to rewards as well as to the opportunity to please their parents.

The fifth message—it is okay to say no, but remember mom and dad are the bosses—is essential for all the skills of positive parenting. Children must always have permission to resist if they are to cooperate. They must be able to resist if they are to make their feelings and wants known to others as well as to themselves. This message strengthens children's willpower, which in turn strengthens their natural will and wish to please and cooperate.

When the five positive messages are the basis of parents' approach to parenting, then the five skills of positive parenting are most effective. In the next five chapters we will explore these messages in greater detail. With this increased insight, parents will be able to make decisions and respond to their children in ways that nurture and support their children in becoming who they truly are and developing the special gifts they have to share in this world.

9

It's Okay to Be Different

Every child is born unique and special. In practical terms, this means that children may be very different from what parents expect them to be. They have their own special gifts, and they have their own unique challenges. To meet their challenges, they will have their own special needs. As parents, our job is not only to tolerate differences, but also to embrace them. This is most effectively accomplished when we are able to recognize what each child's special needs are and to fulfill those needs.

The absence of this positive message is, "Something is wrong with my child. He or she needs to be fixed rather than nurtured," or "My child is bad and needs improvement in some way." Having such an attitude is one of the biggest mistakes parents make. Children need a clear message that they are okay and that differences are fine and to be expected.

The absence of acceptance
manifests itself in the statement:
"Something is wrong with my child."

Applying the five skills of positive parenting makes this acceptance much easier. It is usually when parents are not getting the cooperation they need that they begin to think their children are bad or something is wrong with them. With a greater awareness of how children are different, a parent is not so quick to assume the worst when those differences show up. Rather than resist the differences, parents can nurture children in ways to bring out their unique gifts and strengths as well as to assist them in overcoming their weaknesses.

Every child is a unique combination of different characteristics determined by gender, body type, temperament, personality, intelligence, and style of learning. To be aware of the possible differences, combinations, and permutations of all these factors prepares a parent to accept and embrace the differences. With this expanded insight, it becomes easier to recognize that one child is not better than another, nor is there any one way to be.

**Being different doesn't mean one style is
better than another.**

Often parents mistakenly assume that they know what is best for their children. Even if a child is an apple tree, they persist in trying to help the child be a good pear tree. This kind of help restricts a child's development. Although children are born with an inner blueprint of who they are and what they are here to do, they need their parents' acceptance, love, support, time, and attention to call forth and nurture their potential.

Parents are not responsible for how children turn out, but they are responsible to do their best to bring out the best in their children. Parents need to remember that every child has a unique journey and purpose in this world. To presume that a parent knows best how their children should turn out is to play God.

Children are from heaven. They have within them the seeds of greatness. It is not for parents to determine children's destiny. Instead, parents are to create the fertile ground for children to grow into who they are supposed to be and not who the parent thinks they should be. This special support and acceptance of differences empowers children with the strength and confidence to make their dreams come true.

GENDER DIFFERENCES

Gender differences show up more strongly at adolescence, but clearly, from day one, boys will be boys and girls will be girls. Every child, regardless of gender, has his or her own unique balance of male and female characteristics. Acceptance is important.

Quite often, a mother or father will tend to assume that what is right for her or him is right for their child. This is a mistake. By recognizing common gender differences, it becomes easier to accept and respect certain behaviors and needs that seem foreign. We should not assume that what works for us will always work for our children.

A lack of understanding gender differences can also prevent mothers from appreciating what their mates have to offer and vice versa. Quite often, the mother will instinctively know what is best for a girl, but not for a boy. A father will instinctively know what is best for a boy, but not for a girl. This is because we tend to give our children what we would want or need and not necessarily what they need.

Unless educated about differences, people commonly

assume that others should react and behave the way they do. With an awareness of possible differences, we don't immediately assume that something is wrong when others don't react or respond to life the way we would.

DIFFERENT NEEDS FOR TRUST AND CARING

Boys in general will have special needs that are not as important for girls. Likewise, girls will have needs that may not be that important for boys. Of course, the most important need is love. But love is shown in many different ways. A parent demonstrates love primarily by caring and trusting.

Caring is a willingness to be there for your children, an interest in their well being as well as in who they are, a desire for their happiness and empathy for their pain. Caring is the in-your-face kind of love.

Caring motivates parents to be involved,
interested, and affected by children's
experiences of life.

Trust is a recognition that everything is okay; it is an awareness and belief in your children's ability to succeed and learn from their mistakes; it is an open willingness to let things unfold assuming that everything will be okay. Trust assumes that your child is always doing his or her best even when it doesn't look that way. It gives freedom and space for children to do for themselves.

Trust motivates a parent to give freedom and
space for children to do for themselves.

Certainly, every child needs caring and trust, but in different doses. Too much of a good thing is too much. Up until the age of nine, all children need more caring and a little less trust. After the age of nine, children naturally begin to pull away and become more independent. You can tell a child needs to pull away when he or she starts feeling embarrassed by your behaviors.

Around the age of nine, children begin to develop a sense of self as separate from the parent. This is the time of self-consciousness. From this time on up to age eighteen, children have a greater need for trust, although caring is still important.

Regardless of age, boys need trust more while girls need caring. A boy tends to feel good about himself when he does something on his own. When he can take credit, he feels more confident and proud. For example, he may willfully resist his mother's help in tying his shoes, so that he can get credit and assume responsibility for himself. On the other hand, a girl may feel more loved if you offer to help. Offering to help is a gesture of caring, while letting a boy do it himself is a gesture of trust.

**Regardless of age, boys tend to need trust
more while girls need caring.**

When a mother is too caring for a boy's particular need, he easily interprets her behavior as an indication that "she doesn't trust me to do it myself." When a father is too trusting of a girl's ability to handle things, she can feel that he is not caring enough. When a girl gets too much space, she may feel rejected, hurt, or abandoned. A boy, however, may thrive feeling that his parents recognize his competence and trust in his ability to take care of himself or to do the right things.

Mothers often weaken their sons by worrying too much or

smothering them with concern, while fathers often neglect their daughters' need for caring and attention by giving lots of space, trusting girls to handle things on their own. Parents need to understand that boys form a positive sense of self based on the trust they get, while girls develop a positive sense of self based on the interest and caring attention they get in the relationship.

CONTINUING TO TRUST AND CARE

The biggest challenge in life for women is to trust again after they have been hurt, while for men it is to remain motivated or to continue caring. In response to difficulties in a relationship, women most often complain, "I don't get what I need" (that is, "I can't trust him to give me what I need"), while men complain that "nothing I do makes her happy so why bother" (that is, "I just don't care anymore"). Women most often complain, "He doesn't care any more," and men complain, "She is too hard to please, so I stopped caring."

These different tendencies begin in childhood. Girls and boys come into this world equally trusting and caring. As they experience neglect or the pain of unmet needs and wants, boys often react by caring less, while girls react by trusting less. The challenge for parents is to give a girl extra doses of caring, understanding, and respect so they may continue trusting. On the other hand, the challenge for parents is to give a boy extra doses of trust, acceptance, and appreciation to keep him motivated.

The challenge for parents is to give a girl
extra doses of caring, understanding, and
respect so she may continue trusting.

A girl has a greater need to feel that she can trust her parents to be there and understand her feelings, wishes, and needs. This is her need to be vulnerable and dependent on others. She needs to feel safe in depending on her parents for support. This need is often met by sharing feelings and asking for help. When she is in pain, she needs to know that her parents will be there for her with lots of caring. When she gets the caring she requires, then she can feel trust and remain open. A trusting girl is a happy and fulfilled girl. Safety is essential for a girl to develop her gifts and talents. Otherwise, she feels unworthy, resistant to support, and unlovable.

Sometimes if she feels powerless to get what she needs, a girl may suppress her feminine vulnerabilities and become more like a boy, needing more space, trust, acceptance, and appreciation. For this girl, it is too painful to need the caring and not get it, so she denies her female side and her male side emerges with its needs.

> When a girl is neglected, it is often too
> painful to continue needing and in reaction
> she becomes more masculine.

This does not mean that a girl with more masculine traits is always wounded on her female side. It could also be that she has an active temperament, which may make her appear more masculine. Though they behave more like boys, tomboys are still girls. They still need extra caring, understanding, and respect.

Certainly, a boy needs caring, understanding, and respect to feel safe and trusting, but more important for him is motivation. He needs to be motivated, otherwise he stops caring. When a boy stops caring, he becomes bored, unmanageable, and may have learning problems. When he is not motivated, he will lose his focus and either become depressed or hyperactive. A boy has a greater need to be motivated.

The challenge for parents regarding a boy is to give extra doses of trust, acceptance, and appreciation to keep him motivated.

For a boy to care, he needs to be motivated by success and rewards. He needs to receive the clear message that he can and does make his parents happy. When he is successful in making his parents happy, he continues to be motivated, otherwise he becomes weak and uncaring. Positive rewards for right behaviors are clear signals to him that he has succeeded as well.

While offering help to a girl may make her feel special and cared for, a boy may take it as an insult. Offering to help him may imply that you don't trust him to do what is required. Sometimes the most caring thing you can do for a boy is to give him lots of space to do something on his own. Even if that means he will fail, trust that he will learn his lesson. And please remember, if he does fail, don't tell him, "I told you so."

Offering help to girl may make her feel cared for, but a boy may take it as an insult.

Of course, a girl needs to feel trusted, accepted, and appreciated as well, but, for a little boy to be motivated, he often needs much bigger doses of these. A boy cares more when he is viewed as competent and acceptable just the way he is. Megadoses of trust make a boy feel competent. The "super fuel" to be motivated is appreciation. When a boy feels acknowledged for what he does, he is motivated to do more. There is no greater motivator than success itself.

BOYS ARE FROM MARS, GIRLS ARE FROM VENUS

Understanding that boys have different needs helps parents (especially mothers) make the correct adjustments in giving them what they need. Likewise, by understanding a girl's special needs, parents (especially fathers) make the correct adjustments in giving their daughter what she needs. It is not enough just to love our children and give them what we would want or need most, we must adjust our gifts of loving support to meet their particular needs. Remembering that boys (like men) are from Mars, and girls (like women) are from Venus makes parenting a whole lot easier.

Sometimes if a boy feels powerless to get the trust, acceptance, and appreciation he needs, he may suppress his more masculine characteristics and vulnerabilities and become more like a girl, needing to feel cared for, understood, and respected. For this boy, it is too painful to continue needing trust and not getting it, so he denies his masculine side and his female side emerges with its needs. When he is smothered with caring, he may react by becoming more needy; wanting to feel cared for instead of needing space.

When a boy is smothered with caring,
he may react by becoming more needy.

This doesn't mean that a boy with more feminine traits is always wounded on his male side. It could also be that he has a more sensitive temperament, and in many ways appears more feminine. Sensitive boys often have more feminine hormones and lower levels of male hormones, so naturally they express more feminine tendencies.

Some research has shown that gay men, gifted men, and many left-handed men have significant brain differences from other men. Their brains, like most women's brains, will tend to have billions more neural connectors between the two brain hemispheres. These differences in the brain, coupled with hormone differences, are partially responsible for the making some boys more sensitive. Although these more sensitive boys have more feminine attributes, they are still boys and they still need extra trust, acceptance, and appreciation.

Here are some simple points to jog your memory to remember boys are from Mars and girls are from Venus:

Boys Are from Mars	*Girls Are from Venus*
Boys need more love, attention, and acknowledgment regarding what they do, their ability to do it without help, and the difference they make.	Girls need more love, attention, and acknowledgment regarding who they are, what they feel, and what they want.
Boys need to be admired for what they do more. Acknowledge what he does.	Girls need to be cherished more for who they are. Praise who she is.
Boys have a greater need to be motivated and encouraged.	Girls have a greater need for your assistance and reassurance.
A boy or man is happiest when he feels that he is needed and can provide the support that is needed. He becomes depressed when he feels that he is not needed or he is incompetent to complete the task ahead of him.	A girl or woman is happiest when she feels that she can get the support she needs. She becomes depressed when she feels that she can't get the support she needs and has to do everything herself.

Boys primarily need trust, acceptance, and appreciation in order to be caring and motivated.	Girls primarily need caring, understanding, and respect in order to be trusting and assertive.

MR. FIX-IT

The most common mistake fathers make is to offer solutions instead of empathy when their children are upset and need to express their resistance to life. Men love to solve problems and often pride themselves on being a "Mr. Fix-It." Fathers don't remember that sometimes children just want someone to understand why they are upset rather than to be offered a solution to feel better right away. When children always get solutions, they eventually stop sharing their inner world.

On Mars, they talk about problems when they are looking for solutions, otherwise their attitude is to not talk. "If there is nothing you can do about it, then just forget it." On Venus it is the opposite. Their attitude is: "If there is nothing you can do, then at least we can talk about it." Men generally don't understand or even comprehend that women can get great pleasure from sharing their pain. On Mars, it may seem inexplicable, but on Venus it is a common experience.

In a similar manner, fathers tend to ignore their children's problems by offering solutions or making light of them, not realizing that they now feel put down or minimized. Once my daughter explained why she didn't like being helped with math homework by one of my friends. She said, "Whenever I have a problem, he says, 'That's simple.' It makes me feel like I am stupid for not knowing."

When parents don't sympathize or listen to their children's resistance to life, children misinterpret our intent. When parents have easy solutions, children may feel as if something is wrong with them or they are making too big a deal out of something, rather than feeling safe and nurtured. Before children even consider how upset they should be, they should first feel safe to experience their emotions. When parents restrain themselves from offering quick fixes, children get the trust and caring they need.

Here are some things a father may say that may invalidate children's vulnerable feelings:

Don't worry about it.

It's no big deal.

So what's the point?

This is not that difficult.

It's not so bad.

These things happen.

That's ridiculous.

This is what you should do.

Just do something else.

Just do it.

I don't get it.

Get to the point.

It'll be okay.

It's not so important.

Just deal with it.

What do you want me to do?

Why are you telling me?

With a greater awareness of how they may unknowingly invalidate their children's feelings, fathers can more effectively give the support that girls and boys need. Although women can relate to wanting their husbands to listen, they often forget to listen to their children sometimes. Instead of giving children room to be upset or disappointed, they too will try to fix it.

It is fine to be a problem solver when that is what your children are asking for. In most cases, a parent needs to listen longer and say less to have their children share more and listen more. By giving up trying to solve your children's problems, your job will be easier, and your children will be happier.

MRS. HOME IMPROVEMENT

The most common mistake mothers make is to offer unsolicited advice when children misbehave, make mistakes, or appear to need help. Women love to improve things in life and around the home. It is not that men don't want to improve things, but a man's attitude is "fix it only when it breaks, otherwise leave it alone."

Women realize that no matter how good it
gets, it can always get better.

When a woman loves a man, her tendency to become "Mrs. Home Improvement" gets focused on him. He often resists her unsolicited questions and advice. When a woman becomes a mother, she then focuses her home improvement tendencies on her children. She needs to remember that just as children don't need to be fixed, they also don't need to be improved.

When a mother worries too much or offers too much advice, it smothers children with caring and deprives them of the trust they need. Boys particularly are more harmed by a mother's tendency to worry, correct, and give advice. A good rule of thumb is, for every correction, make sure you have caught and acknowledged your child for doing something right three times. Three positives to one negative is a good ratio.

For every correction, catch your child doing something right three times.

Even better than directly correcting children with advice is simply to direct them into the correct behavior. Instead of saying, "You should be nice to your sister," say instead, "Would you be nice to your sister? I want you both to get along."

By giving children a new direction, you are focusing on success and not on what they did wrong. By focusing on what you want and the opportunity to do that, children's resistance is lessened. Then when the child is ready for explanations, he or she will ask and be receptive.

These are some examples:

You left your plate on the table.	Would you bring your plate over to the sink?
Don't yell in the house.	Please use your inside voice, or (for older kids), Please don't yell.
Your room is still a mess.	Would you please clean up your room?
Your shoes are untied.	Would you please tie your shoelaces?

I've been waiting here for thirty minutes. If you are going to be late, send me a message or call.	If you know you are going to be late, would you send me a message or give me call? I have been waiting thirty minutes.
If you were more organized, you would not have forgotten.	Please take some extra time to get organized and then maybe you won't forget.

Using the five skills of positive parenting to create cooperation frees women from the need to lecture or correct their children. Children naturally learn what is right and good by successfully doing what they are asked to do.

When a mother corrects a child or gives unsolicited advice, the message the child receives is that he or she is not good enough or something is wrong with him or her. The child will feel cared for, but will not feel trusted. As an adult, this child may feel loved by his mother, but not understand why he or she has so much fear or lack of confidence to take risks.

WHEN ADVICE IS GOOD

It is not that advice is wrong. When children are clearly asking for advice, then it is very helpful. The big problem is that mothers give too much advice and, as a result, their children stop listening. It is particularly counterproductive to give good advice when your child is resisting. This means that the child will gradually build up walls against asking for advice when he or she needs it. Giving advice is good when a child is asking for it. If you don't smother children with advice, they will ask for it more as they get older.

Boys are more sensitive to getting solutions than girls. A girl will resist more and continue to share herself, while a boy loses all motivation. When a father or mother gives a boy unsolicited advice, he stops sharing his problems, stops asking questions, and, even more important, stops listening.

Too much fixing makes a girl feel it's unsafe to share, while too much improving makes a boy resistant to listening.

Mothers seek to give advice so that their children don't suffer the same problems over and over. This well-meaning support just shuts a boy down. Then the mother's biggest complaint is that "he won't tell me anything" and "he won't listen." Mothers need to trust more that their children can and will learn on their own or that they will ask.

BOYS FORGET AND GIRLS REMEMBER

A big difference between boys and girls is that boys forget and girls remember. Often a mother becomes overly frustrated because she expects a boy to remember things she has asked. A father often becomes frustrated because his daughter will tend to talk more about problems than he thinks is really necessary. Let's explore why these differences commonly show up.

Men and boys deal with stress by becoming more focused on one thing: one big problem to be solved or one big task at hand. The more stress they have, the more they tend to forget everything but the task at hand. A man can be so focused at work that he easily forgets that it is his birthday, his anniversary, or even his child's birthday.

Under stress, boys become more focused,
while girls need to talk more.

Women often misunderstand this difference and misinterpret a man's forgetfulness as not caring. When she is
stressed, she is inclined to remember more. It is hard for a
woman to forget important things and responsibilities when
she is under stress. This is why, after a stressful day, a
woman often wants to remember and talk about her day,
while a man would rather forget all responsibilities and
watch TV or read the newspaper.

This kind of focused activity is most relaxing for him,
while a woman wants to expand, talk about her day, remember the details, and then let go. A man lets go by forgetting
what was stressful, while a woman lets go by remembering.

This basic difference explains why men and women misunderstand each other so much of the time. Understanding
this difference not only makes our relationships easier, but
also helps us to understand and support our children better.

Understanding differences helps us to
understand and support our children better.

Much of the time, when a little girl appears to be complaining, she really needs time to remember and talk about
her day. This helps a father to understand why he should not
just wait for the point and then give a solution. A girl needs
time, attention, and her father's focus on each word. By giving her his full attention instead of just pretending to listen,
she will get her need satisfied.

A girl literally needs her father's full attention to get

through and release the stress of her day. In applying the skills of positive parenting, parents need to make sure they don't jump to rewards or giving a time out. A girl needs more time to share and express her resistance. Talking is one of the best ways a girl releases stress of the day.

Often, when a little boy forgets to do what a mother has asked, the mother feels he is just not listening. In many cases he has listened, but he has then forgotten. When a boy is stressed, he tends to block out all stressful messages. When a mother makes demands or nags a boy, this is a stressful message, and so he tends to forget it.

When mothers use upset emotions to back up their demands for obedience, these stressful messages are literally forgotten. A mother can greatly benefit from this insight. To help her son remember her requests, she needs to frame them in positive ways. If she leaves out the negative emotions and makes positive requests rather than demands, her son is more likely to remember and respond. Up to the age of nine, when a boy forgets, it is never his fault. He should be expected to forget at times and particularly when he receives stressful messages nagging him to remember.

DIFFERENT GENERATIONS

Every generation is different from the previous one. When parents foster an attitude that embraces differences, children, as they become teens, will not anticipate being rejected for having different ways of thinking about things. Many people today mistakenly think that today's problems result from children being too free. Certainly, this is part of the problem, but taking away freedom is not the solution. The solution is to strengthen the bond between parent and child by using positive-parenting skills.

Taking away freedom is not the solution, but
strengthening the bond of communication is.

Being different does not mean that one is better than the
other. When parents are open-minded about the teen generation, teenagers don't feel they have to pull away to get the
acceptance they need. Even if a parent is very loving and
attentive to their children, if the parent is narrow-minded,
the teenager often feels an urge to oppose and rebel; to
break out of their narrow limits. If you hold your values of
what is good but do not condemn others, your teens will feel
it's safe to come to you for support. Otherwise, they will
break those lines of communication.

THE CULTURE OF VIOLENCE

Today more than ever teenagers need clear and open lines of
communication with their parents. The challenges that our
teens face are enormous. Without parental support, it is
extremely difficult not to be swayed by negative influences.
Teens are already vulnerable to peer pressure. If they don't
have a strong foundation of positive communication with
their parents, it is very hard to stay connected to who they
are and to hold on to their own values and wants.

Without this anchor of parental communication, a teen
is easily tossed around by the dangerously high waves of
negativity in the world. Teens and even preteens can be very
mean. Without strong support from home, children will easily succumb to peer pressure to experiment with drugs,
drinking, violence, gangs, stealing, lying, cheating, and sexual promiscuity in order to gain acceptance. When teenagers

do not feel accepted at home, they are willing to give up their values to seek acceptance from their peers.

When teenagers do not feel accepted at
home, they seek acceptance from their peers.

Today our teens enter a culture of violence. They are more sensitive than any previous generation. This means that what goes in comes right back out. When teenagers are not sensitive or open, they are not affected by the outer world as much. In a free society with so many choices, our children are much more vulnerable to being influenced by others. One bad apple does spoil a whole barrel of apples.

On one hand, our teenagers feel a healthy need to be more independent, and on the other, they need our support more than ever. To give this support effectively, parents have to back off from fixing and improving and instead be an open-minded resource so that our children want our positive support.

When expressing our opinions, we must also be careful to give our children support for holding different opinions. When parents insist on "one-way thinking," their teens will insist on the other way. Be open-minded and your children will be free to make their choices instead of just reacting or rebelling against yours. When children grow up in an environment that accepts differences, they will not feel so pressured to be like their peers. They will assume and assert their right to be strong-willed and different.

Be open-minded and your children will be free
to make their choices instead of just rebelling.

To support our children, we must hold back advice, rigid judgments, and solutions in order to keep the lines of communication open. Fortunately, it is never too late to open these lines. Using positive-parenting communication skills and applying the five positive messages can begin to open up those lines of communication at any age.

DIFFERENT TEMPERAMENTS

As we have explored in Chapter 4, there are basically four temperaments: sensitive, active, responsive, and receptive.

1. Sensitive children have stronger feelings, go deeper, and are more serious.

2. Active children have strong wills, take risks, and want to be the center of attention.

3. Responsive children are bright, light, and need more stimulation; they move from one thing to another.

4. Receptive children are well mannered and cooperative; they follow instructions well but resist change.

Although most children have at least a little of each temperament, generally one or two of these temperaments is dominant. With an understanding of how temperaments differ, parents can easily identify their child's predominant temperament and learn what that child needs. (Refer to Chapter 4 to identify the needs and particular skills required to nurture a temperament.)

When a child's temperament is different from that of the parents, unless the parents are aware of all four temperaments, it is very difficult to nurture this child. So much unnecessary wounding and neglect occurs because parents are not educated with an understanding of these simple and basic differences.

Unless parents are aware of all four temperaments, it is very difficult to nurture a child whose temperament differs from theirs.

Quite often, some of the biggest problems parents have in getting along is disagreeing about what their children need. A *receptive* parent will instinctively know what a *receptive* child needs, but if the other parent is *active, sensitive, or responsive* he or she cannot instinctively know what the child needs. As parents, we cannot always assume that what works for us will work for our children. Not only does the child suffer, but the parents argue needlessly.

For example, without an understanding of different temperaments, a *responsive* parent would not only think something is wrong with a *receptive* child's resistance to change, but could not give the child the rhythm and repetition that *receptive* children need.

On the other hand, a *receptive* parent, who doesn't like change but likes repetition, would think something is wrong with the *responsive* child who never finishes things. Without this important awareness, the parent would not give the child the varied activities that he or she needs.

HOW TEMPERAMENTS TRANSFORM

When parents learn how to accept and nurture the different temperaments, they naturally transform and flower. Some children may start out with a little of all four types and gradually move through them all throughout their lives. When a temperament is nurtured, at least for a while, it will transform into the next. These are some of the transformations to be expected:

Sensitive children, who have stronger feelings, go deeper, and are more serious, gradually lighten up, and have lots of fun and laughter along with being original. Sensitive children become more responsive. When a serious child feels heard, he or she will tend to become light and cheery for a while.

Responsive children, who are bright, light, and need more stimulation while moving from one thing to another, gradually learn to focus, be disciplined, and fully commit themselves in relationships and work. Responsive children become more receptive. When responsive children get to do many things, they begin to find something they really like and become more focused for a while.

Receptive children, who are well mannered and cooperative, and follow instructions well, but resist change, gradually become self-motivated, wise, adaptive, and flexible. Receptive children become more active. When receptive children have a regular routine, they feel safe enough to take risks and try new things.

Active children, who have strong wills, take risks, and want to be the center of attention, gradually become cooperative and compassionate in service to others. Active children become more sensitive. When active children get enough structure and guidance to feel competent or successful in achieving their goals, they become more sensitive and aware of the needs of others and wish to serve.

AFTERNOON ACTIVITIES

Based on these different temperaments, we can see better what activities are most appropriate for a child. Keeping temperaments in mind, let's explore afternoon activities.

The sensitive child needs lots of understanding.

It is hard for sensitive children to start new friendships, so they need a little extra help. Parents need to put a sensitive child in a supervised activity that promotes safe, harmonious interactions. This child doesn't need a lot of stimulation; too much is definitely too much. Sensitive children need to be around people who have similar abilities and sensitivities. It is especially good for them to help in caring for a pet. A pet or stuffed animal always understands what they are going through.

The responsive child needs a greater degree of variety in activities than other children.

Parents who give their children lots of stimulation in the afternoon nurture this important need. Camps, museums, parks, malls, sports, gymnastics, skating, movies, some TV, video games, books, walks, swimming, swings—all give stimulation. These children can easily become addicted to video games or TV and become distressed inside because they are not getting other kinds of natural stimulation.

The receptive child needs a regular routine each day.

Too many activities disturb a receptive child's rhythm. He or she could come home every day and read, walk the dog, watch a few TV shows, snack, and do some homework. Receptive children thrive on routine and don't like lots of change. Being around responsive or active siblings too much can create distress. They like to watch action, but when required to participate too much, they become stressed. If they are left in day care or involved in after-school activities, teachers should be alerted that these children have the right to watch and should not always be put on the spot to get into the action.

The active child needs lots of structure.

Active children need lots of supervision, rules, leaders, and action. Supervised sports and teams are great for these kids. Leave these children on their own and they will become bossy, get into trouble, and lead others into trouble as well. When there is one bad apple who tends to bring everyone else down, this is the one.

DIFFERENT BODY TYPES

Another area of difference that parents need to understand is that some children have a body type that differs from their parents', and all types should be appreciated equally. This can be very difficult because body types go in and out of style. In countries and times where people have little food, fat is considered beautiful. Yet, where food is in abundance, thin is beautiful. Regardless of fashion, muscles in men are always in fashion.

In countries and times where people have less food, fat is considered beautiful.

Regardless of fashion or the current social view of the body, children are born with specific body types that do not change much. There are thin or rectangular people, fat or round people, and muscular or triangular people. Every person is born with a body type and it doesn't change much. While there are three basic body types, there can be millions of combinations and permutations.

Sometimes children who are fat do become thinner or more muscular; people who are muscular become fatter or thinner; and thin people become more muscular or fatter. The important message is acceptance. Each child is different. If everyone looked the same, the world would be very boring. To expect a round child to be thin is unrealistic. So many girls and boys feel inadequate because their parents are obsessing about their weight, or they have just given up and don't care about how they look.

To expect a round child to be thin is unrealistic.

To be a good role model, parents need to be accepting of their own bodies and be vigilant about staying at a healthy weight. For most people, that means accepting that you are not going to look like a model. Just as girls tend to have weight issues, some boys have muscle issues. They are not as big or strong as others and wonder why their muscles don't get so big.

A mother or father needs to explain to their child that every person is unique and different. Muscular bodies respond differently to exercise than thin ones. In a similar way, a parent could explain to a child with weight issues that some people eat more food and don't put on weight, whereas others need to be a bit more careful. Otherwise, a round child mistakenly assumes that they eat too much or that they have no discipline to eat the right amount.

DIFFERENT INTELLIGENCE

Another area of difference is in intelligence. To be most supportive and appreciative of your child's gifts, it is important to realize that there are different kinds of intelligence. In the West, we have become too focused on the model that intelligence is measured by an IQ test. Ironically, these tests are arbitrary and tend to make boys' IQs seem higher and girls' IQs lower. When IQ test problems focus on spatial abilities, boys score higher; and when test problems focus on language skills, then girls score higher.

Besides discriminating against girls, tests that measure intelligence do not consider all kinds of intelligence. Someone decides what problems are on the test and therefore determines the outcome. IQ tests only measure a certain kind of intelligence and in no way is having a high IQ linked to success in life, relationships, or work or a low IQ to failure in these areas.

Besides discriminating against girls,
IQ tests do not consider different kinds
of intelligence.

With so many out-of-work and divorced Ph.D.s, it is now becoming common knowledge that academic success does not ensure job success, or life success for that matter. Children who have a higher degree of academic intelligence do better in public schools as they are set up today, but that in no way ensures success in life, work, or relationships.

Unfortunately, children who have other kinds of intelligence are not recognized and nurtured as much in public schools. There are basically eight kinds of intelligence and every child is born with a unique distribution of each. Each of the types of intelligence are like different colors we can use to paint the landscape of our lives. They include: academic, emotional, physical, creative, artistic, common sense, intuitive, and gifted intelligence. Every child is born with different degrees of each intelligence, and each type can be stimulated to higher degrees of development with the right kind of nurturing.

Academic Intelligence

Children who have strong academic intelligence do well in school. They can sit, listen, and learn. They are able to absorb, comprehend, and repeat the knowledge they are taught. If the knowledge is presented to them, they are able to remember it. This does not necessarily mean that they can apply the knowledge or use it constructively in life.

Adults know that much of what they learned in school is forgotten, but school does teach us to think, analyze, comprehend, and find resources. Academic intelligence is stimulated by reading, writing, and listening to lectures. Parents need to give these children academic opportunities.

Emotional Intelligence

Children with strong emotional intelligence are able to create and maintain healthy relationships with others and themselves. They are more aware of how others think and feel and can empathize with another person's point of view. This ability to connect and be compassionate serves them well, not only in their personal lives, but also in the work world.

Successful people in the work world must have a high level of emotional intelligence. This intelligence also gives us the ability to manage and articulate our inner feelings, wishes, and wants. More and more schools are including programs for understanding feelings, developing empathy, and improving interpersonal communication. Parents need to give these children opportunities for social interaction and must themselves have good communication skills.

Physical Intelligence

Children with strong physical intelligence easily do well at sports and are able to keep their bodies strong, healthy, and vital. They instinctively understand their body's need for exercise and good food. To develop their athletic abilities, they need opportunities for practice and coaching. Their innate abilities can be dramatically improved by having opportunities to compete with other children. Healthy competition draws out the best in them. They need positive acknowledgment to develop self-esteem. They not only feel good, but also know how to look good. Physical intelligence extends beyond sports to the health of their body. They need to know about their body and what makes them strong and vibrant. Their positive appearance and vitality helps them to be successful in the world.

Creative Intelligence

Children with creative intelligence have a more developed sense of imagination. They can play games with a few blocks or faceless dolls. They often create imaginary friends. They don't need a lot to be stimulated. When too much is done for them, they don't develop their imagination. They respond well to listening to stories, because they are required to use their imagination to create the scenes and characters.

Too much TV, where the images are visual, can weaken children's ability to imagine. Just as every intelligence grows by being used, a creative intelligence grows when imagination is stimulated, enabling children to think differently. They succeed in life where others fail because they can look at things in a new and different way.

Many successful entrepreneurs didn't have a formal education or didn't do well in school, but succeeded because they were creative. While growing up, they often received support for thinking differently. They were empowered to create their niche in life. They tend to be more original and succeed by doing their own thing. They are often left-handed. Parents need to give these children a lot of support for thinking differently and solving problems.

Artistic Intelligence

Children with artistic intelligence are naturally more interested in singing, drawing, design, writing, acting, drama, comedy, and other forms of artistic expression. They need to be stimulated by others who have already mastered their artistic talents. Though all children need role models, these children particularly need accomplished role models of artistic intelligence. These children are different, more sensitive, and often don't get the emotional support they need.

Parents need to encourage these children to follow their dreams and develop their unique talents and artistic abilities. For artistic intelligence to flourish, children need role models and opportunities to practice and develop their intelligence with lots of encouragement and appreciation from their parents.

Common Sense Intelligence

Children with common sense intelligence are often bored by intellectual lectures. They just want practical information. This intelligence is on the rise in the West. There is so much information available that people just want what is necessary. These children focus on what is useful to them and will often challenge what is taught in school as not being relevant to their lives.

To keep children interested, many schools are trying to update their programs to keep them timely and relevant. Children with common sense intelligence need basic skills to use in their lives, relationships, and work. They are not motivated to memorize information unless it has a functional value.

Common sense intelligence allows a person to live a stable and grounded life. They are not easily swayed by lofty ideas that are no longer relevant to the world today. They are eager to apply what works for them. They need opportunities to put into practice what they know and learn by doing and evaluating the results. This intelligence develops by giving your child structured activities with lots of freedom and independence.

Intuitive Intelligence

Children with strong intuition just know things. They don't need to be taught or told. Information simply comes to them. It can be information in a subject of study or information another person has. They tend to be more spiritually inclined. All they have to do is to read a few sentences of a book and they intuitively get much of the content. Besides intuitively sensing the content, they get the benefit of knowing the content.

For example, if you were to read a book on social skills, in future situations that content would be background information to help you to respond appropriately to situations. You would have a stronger feel of what to do. This is the benefit of having read that book. Children with intuitive intelligence can benefit from a teacher's knowledge without having to study all the details.

Children with intuitive intelligence are often discounted. Most parents or schools don't have programs for developing this intelligence. For intuitive children, parents need to worry less about academic performance and appreciate their sixth sense for knowing things that are needed. This kind of intuitive intelligence is primarily stimulated by personal contact and not through TV programs, computers, or books.

Gifted Intelligence

Children with gifted intelligence tend to be particularly good at certain kinds of intelligence, but low in others. All children are born with great intelligence, but in different proportions. Gifted children get a lot of one, but little in others.

To lead a happy and fulfilled life, gifted children need special support and guidance to challenge their special gifts, otherwise they will become bored and unmotivated. In addition, gifted children need special support to develop the skills and intelligence areas in which they are weak.

People who are especially brilliant in one field often suffer in their lives, because their other kinds of intelligence were not nurtured. A brilliant scientist or billionaire entrepreneur may not be able to say "I love you" to his or her spouse. Many people are emotionally gifted, but have poor health. These loving people care for others, but don't take care of their own body with regular exercise. Traditionally, great artists have struggled in life because they were missing the common sense required to manage money and other more mundane aspects of life. There are endless examples of great gifted people who suffered enormously in their lives.

Some people are gifted with enormous physical intelligence. They always look great. They are so used to getting love and support for looking great that they are afraid to reveal more of who they are inside and lose all that immediate attention and adoration. This is why "beautiful people" are sometimes very superficial. Their development is stunted because they don't want to risk losing the love they get by showing up and looking the way they look.

**The risk of failure may hold children back
from learning new skills.**

This same principal is true for all the different kinds of intelligence. For example, people who are academically inclined are sometimes weaker in social skills. They enjoy being excellent in one field. They get a lot of love and attention for being the best at something. To try working and developing another weaker intelligence is too big a risk.

The thinking here is simple. If I am better at something, then I get love and support. If I am not better, then I will lose love and support. To counter this thinking, these children need encouragement to develop other areas of intelligence in which they are not gifted. In this process, they learn through experience that they don't have to be better or the best to be loved. As a result, they are able to lead more balanced, fulfilled, and successful lives.

DIFFERENT SPEEDS OF LEARNING

An old adage from Shakespeare says, "Some are born great, some achieve greatness, and some have greatest thrust upon 'em." This simple truth, when combined with an awareness of the different kinds of intelligence, helps parents to understand and respect how their child learns differently.

Children may be gifted or "born great" with one or two kinds of intelligence. They may be gradual learners or "achieve greatness" with a few more kinds of intelligence. With the other kinds of intelligence, they may be late bloomers and "have greatness thrust upon 'em."

People exhibiting these three ways of learning can easily be summarized as: runners, walkers, and jumpers. To explore these three different learning rates in greater detail, let's take learning to ride a bike as an example:

Runners

This child sees another child riding a bike and just gets on and rides off. Children with this learning style are runners. They are fast learners, but to stay interested and involved, they need to be challenged. They learn very quickly, because they are generally gifted at what they are learning. Parents must be careful to make sure that runners get the opportunity to develop their other kinds of intelligence that may not be so easy for them.

Walkers

This child takes a few weeks to learn how to ride a bike. These children respond well to instruction and with each attempt they get a little better. They may start out with training wheels but within a couple of weeks are riding on their own. Walkers are what parents call "dream children" or "easy." They always learn a little more, get better, and clearly let you know you are helping and they are learning. These children are so easy to manage that they often miss a lot of important nurturing and attention.

Jumpers

This child is the most difficult and challenging for parents. These children may take several years to learn how to ride a bike. They take instruction in, but don't progress. They don't get better, they don't show any signs of learning, and the parent has no idea if anything they are doing is helping. If the parent persists, two years later the child gets on the bike and suddenly rides.

All that instruction was going in, but the parent had no indication of progress. Then, in one mysterious moment, these children somehow put it all together, get on the bike, and ride as if they had been riding for two years. On the surface, it may have looked as if no progress was being made, but then suddenly in one jump they get there. These children often don't get the time and attention they need to make the jump. Without parental encouragement and persistence, they quit and never realize their inner potential.

GOOD HERE BUT NOT GOOD THERE

A child could be a jumper (slow learner) when it comes to riding bikes, but a runner (very fast learner) when it comes to social skills. He or she could be the nicest and most cooperative child while making dinner together or traveling on a trip, but then, when it comes to riding a bike, a change occurs. Immediately, your child becomes resistant and uncooperative. By understanding different learning speeds, a parent can be more patient and accepting of their child's resistance. All children excel at some skills, but resist others. Being good here and not good there is natural and normal.

Just because a child is a jumper and appears to be a slow learner does not mean that he or she has low levels of that intelligence. Sometimes it is those areas where we resist learning the most that we have our greatest strengths. For me personally, I was never a good writer or public speaker and resisted writing and speaking in a group. Both were gifts that were to come much later in life.

On the other hand, just because someone is a runner or walker in a particular area of intelligence doesn't imply that he or she will excel in this field or have a tremendous potential for growth. For example, the majority of people who get a university degree in a particular subject don't follow that particular path later in life. Getting a degree in anthropology doesn't mean that you will be an anthropologist. The easiest path or path of least resistance is not always our greatest strength.

COMPARING CHILDREN

One of the big mistakes parents make is to compare their children to one another. If you have a child who is a walker in most areas of intelligence, then everything is relatively smooth and easy. When your next child is a jumper in some areas and resists more, you may mistakenly assume that something is wrong with the child.

Jumpers never seem to be learning or listening. You teach them to set the table and they forget how. You teach them table manners and they keep forgetting. You teach them their math tables and they keep forgetting. You teach them to speak clearly and they don't speak. You teach them to tie their shoes and they can't. You explain their homework and they just can't get it.

Without positive-parenting skills, these children usually get punished again and again, which makes it even more difficult for them to develop confidence. Children can only grow in confidence when they get the consistent messages that they are not being compared and that they are good enough just the way they are. Every child is unique and special and deserves love just the way he or she is. By understanding all the different ways healthy loving children can be different, it is easier to be an accepting and supportive parent.

Reviewing this chapter from time to time can make the process of parenting dramatically easier. Times of frustration are caused by expecting our children to be different from the way they are. Just remembering that they are supposed to be different helps us to relax and reflect on a more appropriate way of dealing with our children.

10

It's Okay to
Make Mistakes

Besides being unique and different, every child comes into this world with his or her own bundle of issues and problems. No child is perfect. All children make mistakes. *Everyone* makes mistakes. To expect children not to make mistakes gives them a cruel and inaccurate message about life. It sets a standard that can never be lived up to. When parents expect perfection, children can only feel inadequate and powerless to live up to their parents' standards.

**All children make mistakes; it is perfectly
normal and to be expected.**

Parents constantly need to adjust their standards for and expectations of their children according to their natural abilities. At every age, children's abilities change naturally. Every child has different abilities. When weak in a particular area, a child will need more help and sometimes will need the parent to carry him. Children should not get the message that something is wrong with them for making mistakes. Too many shaming messages make children feel they are bad, unworthy, or that something is wrong with them. They feel defeated and lose their natural motivation and confidence.

FROM INNOCENCE TO RESPONSIBILITY

Young children, up to nine years old, are not capable of dealing with shaming messages without assuming too much blame. Any kind of punishment, disapproval, or emotional upset in reaction to your child's mistakes ultimately gives a shaming message. When there is a problem, unless someone else assumes responsibility, the child will assume too much blame.

Before the age of nine, a child cannot discern the difference between I *did* something bad and I *am* bad. Children younger than nine are not capable of logical thinking. A child reacts in this way: "If I did something bad, then I am bad," or "If what I did was not good enough, then I am not good enough."

Without a sense of self, when a child makes a mistake she has nothing to fall back on. If she makes a mistake, she is a mistake. When a child assumes too much responsibility, a parent can correct this tendency by assuming responsibility themselves. When parents assume responsibility for what happens to their children, children don't take it on.

Many adults suffer from low self-esteem and feelings of unworthiness, because they still cannot discern this difference. When they make mistakes, they conclude that they are not good enough. Although these adults are capable of logical thinking, when they were younger than nine, they were not nurtured to experience their inner innocence. They may even reason that they are not bad, but inside they still feel bad or unworthy.

An adult with healthy self-esteem reacts to his or her

mistakes with acceptance and a willingness to learn from those mistakes. These are some examples of the ways a healthy adult reacts logically to mistakes:

If I did something bad, I'm not bad, because I did not know better.

If I did something bad, I am not bad, because I do a lot of good things as well.

If I did something bad, I am not bad, because I can now learn from my mistake and do it better.

If I did something bad, I am not bad, because I can make amends or make it up.

If I did something bad, I am not bad, because I was doing my best. Others make mistakes, and they are not bad.

If I did something bad, I am not bad, because I didn't mean to do it.

If I did something bad, I am not bad, because it was an accident.

If what I did was not good enough, I am still good enough, because I am learning, and soon I will be good enough.

If what I did was not good enough, I am still good enough, because I am not expected to be perfect.

If what I did was not good enough, I am still good enough, because today I was sick, not feeling well, or having a hard day.

If what I did was not good enough, I am still good

enough, because today's challenge was more difficult than usual.

If what I did was not good enough, I am still good enough, because nobody wins every time.

If what I did was not good enough, I am still good enough, because I recognize my mistake, and can correct it in the future.

If what I did was not good enough, I am still good enough, because others can't do it either.

In each of these examples, one is able to use logical thinking to discern the difference between "I did something bad" and "I am bad." Research on child development clearly shows that children younger than nine do not have this capacity for logical thinking. By focusing on your child's mistakes, the child comes to feel that he or she is bad or inadequate. Instead of focusing on the problem, positive parenting focuses more attention on the solution. By learning that they are good, children stay open to your direction with a willingness to cooperate; children who are shamed close down.

Children younger than nine are not capable
of dealing with shaming messages without
assuming too much blame.

Shaming messages are always counterproductive. After the age of nine, it is appropriate to ask children to take responsibility and make up for their mistakes in some way. The first nine years are a time to develop innocence, then, during the following nine years, children learn to become responsible. When a child turns nine, he or she is ready to begin taking more responsibility for mistakes by making amends. When a child is younger than nine, a parent should ignore, overlook, or have a neutral attitude about the child's mistakes.

When children are made to feel responsible for mistakes too soon, they begin to feel bad and unworthy in various ways. Without a strong foundation of innocence, children's natural ability to self-correct doesn't have a chance to develop.

Instead of shaming or punishing to correct your children's behavior, apply the five new skills of positive parenting. Instead of focusing on the problem, just ask for what you want in the future. Don't dwell on the problem—move ahead to the solution.

When a child knocks over a vase, it does no good at all to focus on the mistake. Instead, just say something loving and warm like this, "Oh, this pretty vase is broken. We must be careful around vases; they are delicate and break easily. Let's stop everything and clean up this mess."

**It does no good at all to focus on
a child's mistake.**

Getting upset with the child does not increase his ability to learn from a mistake. It only confuses the child more and interferes with his or her development. When a child breaks a vase, the mother may feel, "I need to make sure he knows this is not okay and he needs to listen to me." The new skills of positive parenting make this shaming message unnecessary. The outcome the mother is really looking for is increased cooperation. The five skills will accomplish that.

It is not even important for the child to acknowledge that he knocked over the vase. Some children, because they fear punishment or disapproval, will deny their involvement. The real problem in this instance is the child's fear of the parent not his denial. Particularly for a child younger than nine years old, it is of no value to put the child on the stand and cross-examine him to prove his guilt. This focuses too much attention on the problem and not the solution. The solution is finding a way to motivate greater cooperation.

The child doesn't need to be punished or lectured to recognize how expensive or precious the vase was. Before age six or seven, children can't even comprehend monetary value. To children, five dollars, five hundred dollars, or five thousand dollars is the same. If the parent applies the five skills of positive parenting, the child will automatically be more careful and considerate in the future, not just regarding the vase, but with everything else as well. Instead of weakening children's willingness to cooperate by shaming or punishing, a wise parent overlooks the mistake with a neutral or bored attitude and focuses on cleaning up after the mistake.

Even with a preteen (ages nine to thirteen) or teenager, it does little good to focus attention on the mistake if she denies it. Preteens and teenagers commonly believe that if you can't prove that they did something, then it is as if they didn't do it. Rather than attempting to prove their guilt, the wise parent recognizes the bigger problem: The teen doesn't feel it's safe to be held accountable.

In this case, the parent can simply explain what would have happened if she had been responsible for the broken vase and then let go. When the teen realizes that she would only be required to clean up the mess and that there would be no great punishment or loss of love, she will be more inclined to feel accountable and be responsible for her mistakes in the future.

WHOSE FAULT IS IT ANYWAY?

From the perspective of positive parenting, when a seven year old knocks over a vase, it is not her fault. She is seven and cannot be expected to understand the value of the vase. Even if she can understand its value, she can't be expected to remember it. When seven year olds play, they sometimes break things. Even if the parent told the child not to touch the vase, it is still not her fault, because the child forgot during that period of time. In a very real sense, she was out of control. When the child loses control, what occurs is not her fault.

If the brakes on a car stop working, it is not your fault that the car goes out of control. It is not due to your mistake that the car crashes, because there was nothing you could do about it. When you are out of control, because the brakes are not working, crashing the car is not your fault.

> When your brakes are not working,
> crashing the car is not your fault.

Yet from a different perspective, if the car's brakes fail, it is your fault. It was your car and you were driving when you smashed into someone's hedge. From the perspective of being responsible to correct the problem, it is your fault. After all, someone has to pay for the mess.

Still it is not so clear-cut. It may be your mechanic's fault for not noticing and correcting the problem. Do you pay or does the car mechanic who just serviced your car pay?

Maybe it is the car dealer's fault because he sold you a lemon. Maybe it is the carmaker's fault because that car model should have been recalled.

From this simple example, it becomes clear that determining fault and responsibility is a complex matter that often requires many expert lawyers and judges to determine. If adults can't handle determining fault without lawyers and judges, how can we expect our children to deal with it?

Fortunately, positive-parenting skills provide a practical alternative to making our children pay for their mistakes in order to motivate cooperation in the future. When children are raised this way, as adults they have a greater sense of responsibility, cooperation, and motivation to correct their mistakes, and there will be less need for lawyers, judges, and courts.

LEARNING RESPONSIBILITY

Instead of teaching our children to feel bad for their mistakes, we need to teach them to learn from their mistakes and, when appropriate, to be responsible to make amends for their mistakes. Some parents are happy to stop punishing or making their children pay for their mistakes, but they worry that their children will not learn to be accountable or responsible. This is an important consideration.

After all, you cannot learn from a mistake or be responsible to make amends unless you first recognize your mistake. This accountability is certainly essential for adults to self-correct, but it is not so for children. Children don't need to be accountable to learn from a mistake. Babies have no sense of self at all, and yet they are constantly learning and self-correcting.

Accountability is essential for adults to self-correct but it is not so for children.

Accountability is the conscious recognition that "I made a mistake." Children do not develop a sense of self until they are nine years old. Before the age of nine, self-correction occurs automatically without accountability. There is no sense of a self who has made the mistake. The innocent child self-corrects, not because he has done something wrong, but to imitate his parents and to cooperate.

When children are held accountable and responsible for their mistakes, it restricts the natural ability to self-correct by means of imitation and cooperation. This self-correction is essential for learning and growing. Life is always a process of trial and error. Everyone makes mistakes.

Those who succeed in life are those who can self-correct and change their thinking, attitude, or behavior.

The sense of accountability and responsibility without feeling unworthy or inadequate comes from feeling comfortable making mistakes. After nine years of feeling safe to make mistakes without punishment or the loss of love, children are ready to assume an appropriate sense of accountability and responsibility. When it is really okay to make mistakes, then it is safe for children to recognize their mistakes and consciously learn from them.

HARDWIRED TO SELF-CORRECT

Children are hardwired to self-correct automatically after making a mistake. The main reason children or adults don't self-correct after making a mistake is that they don't feel safe admitting they made a mistake. This natural self-corrective reaction requires feeling safe to make mistakes.

To the degree to which children feel afraid of making mistakes, they lose the natural ability to self-correct.

Anxiety about making mistakes only increases the chances of making mistakes. Punishing or shaming children for making mistakes only increases the anxiety and weakens the natural ability to self-correct. Parents need to remember that children are from heaven. Self-correction is an automatic process that occurs primarily by means of imitation and cooperation and not punishing and shaming.

YOUR CHILD'S LEARNING CURVE

Even when punishment and shaming is avoided, the process of self-correction is gradual. As with any behavior or attitude, there is a learning curve. As we have already explored, each child's learning curve for a particular task is different. Some learn to ride bikes faster, while others will get ready for bed easily. Even if a parent is doing everything right, the challenge of getting a particular child to behave at the dinner table may take time and require a good deal of energy, effort, and attention.

> When children take longer to learn, it is
> neither their fault nor the parents'; it is just
> what is required.

By simply giving directions and demonstrating correct behavior, a parent teaches a child right behavior resulting from constant self-correction. This is all a parent can do, and the rest is up to the child and will be done in his or her own time.

Parents tend to be very generous with their love and patience in reaction to children's mistakes until they learn to communicate. At this point, parents mistakenly assume the child can hear and understand the reasoning of their request. Expecting a child not to draw on the wall because it won't come off requires logical thinking that a child isn't capable of. Expecting a child to go right to bed because she will feel better tomorrow requires too much reasoning.

When a baby drops its food on the floor, a parent is forgiving and patient because it is clear that the baby does not know better. Once the child can communicate, the parent assumes that the child should know better. Why? Because he or she has been told before, and the parent has given a logical reason. Parents assume that just because a child can communicate, he or she can comprehend the meaning or reason for the request.

In some cases, a child may need to be told two hundred times before learning a behavior. A runner may learn in one request, but a jumper may take two hundred times, before suddenly mastering the behavior. A walker will show progress with each request, but will still take a lot of repetition. Understanding and accepting your child's unique learning curve is essential to provide the loving support your child needs to self-correct.

UNDERSTANDING REPETITION

This concept of repetition is easy to comprehend when we consider learning to hit a baseball. By continuing to swing the bat to hit the ball, it may take two hundred attempts before you can hit that ball where you want it to go. After two hundred learning opportunities, the ball goes where you want it to go.

In a similar manner, it may take two hundred learning opportunities before a child automatically responds to a particular request such as, "Please, don't throw your food on the floor; keep your food on your plate." When feeding a very young child, a mother wisely puts plastic on the floor.

Parents need to remember that, even though a child can communicate, the child's brain is still developing and changing every day. The child is not motivated by reason or logic. To continue self-correcting, he or she is motivated by repeated requests or commands.

A child may consciously appear to learn from a mistake, but doesn't. Up until the age of nine, a child learns by imitation and following directions. If you feel bad about something, your child will feel bad about it, too. Children imitate your reaction, but they don't understand it. They are not capable of logical reasoning until age nine.

If you get upset about children's mistakes, they will feel bad, but that doesn't in any way mean they have learned anything except to fear your reactions and suppress their will. Positive parenting allows you to preserve children's will and strengthen their will to cooperate. This is how children learn best.

LEARNING FROM MISTAKES

One of the biggest mistakes parents make is to assume that children can logically learn from their mistakes before age nine. Parents try to teach children to learn from their mistakes rather than focusing their attention on increasing cooperation and guiding them in appropriate behavior. When children make lots of mistakes, it is no wonder that they just continue to do so.

Children cannot and do not consciously or logically learn from their mistakes before age nine. They do, however, automatically self-correct. They learn behavior skills like respect, offering help, listening, cooperating, and sharing by imitation and direction.

By being guided again and again to do the right thing, children eventually learn what is right.

When children continue to make disturbing mistakes or forget what you asked, it is often because they are not getting the structure, rhythm, or supervision that they need. From this perspective, when a child makes mistakes, the parents are always responsible. It is actually incorrect to make your child feel responsible.

Until children are nine years old, they are not capable of self-supervision or direction, because, without logical thinking, children can only imitate. From ages nine through eighteen, children are still only learning to be accountable and responsible. Once they can be fully responsible, around age nineteen, we set them free to live in a world that holds them fully accountable for their actions.

Many parents mistakenly expect the sense
of responsibility of a nineteen year old from
a young child.

Children learn to be respectful of objects or others not by being held accountable or responsible, but by seeing their parents are respectful. It is not uncommon for a mother to hit her child saying, "I don't want you to hit your brother. Now say you are sorry to him."

Children learn by imitation. If the parents yell or hit their children, then children hit and yell at each other. If parents respect and apologize in relating to each other, then their children eventually respect each other and are better prepared to be accountable and responsible at nine years old. When children are able to feel innocent and not responsible for mistakes for nine years, then they will be prepared to learn from mistakes when their brain has developed sufficiently.

Children who do not experience nine years of innocence are also able to learn from mistakes, but they may not easily forgive themselves for their mistakes. They will have a greater tendency to be defensive about making mistakes and fail to self-correct. When children experience nine years of innocence, they possess a strong foundation to deal with mistakes in a healthy way. They are able to forgive themselves and gradually learn from mistakes.

Before the age of nine, children know what to do, because they have been guided by parental requests. With positive-parenting skills, children are motivated to cooperate not from the fear of punishment but from the natural instinct to please and cooperate.

Focusing on mistakes does not
serve children in any way.

When children cooperate with their parents' will and wish, they learn what is right. Children don't learn what is right by analyzing what they did wrong. It is a mistake to give a child a time out and ask him or her to think about what he or she did wrong. When you give a time out, it is simply to bring a child back into control and not to teach what is right or wrong.

When children cooperate with the parent and gets positive feedback for cooperating, they learn to behave in a responsible manner. They are "able" to be "responsive" to their parents' needs, wishes, and wants. In this way, the child is responsible. This is the only kind of responsibility a child younger that nine can develop.

LEARNING TO MAKE AMENDS

With the safety to make mistakes, children can then focus on what to do after they make a mistake. When a mistake is made, parents need to demonstrate how to make things right or better again. By imitating their parents behavior, children gradually learn how to set things right again or how to make amends.

The best way to teach this important lesson is to lead by example. If your son should hurt a friend while wrestling, take your son's hand and go over to the friend who is hurt. While your son listens, say, "I am sorry this happened. Let's make it better." With your son's assistance, get some ice and take time tending to the injured child's wound. Rather than blame your son for a mistake he made, share a sense of loss with him.

Demonstrate to your child that there is always the chance in life to re-balance the scales when a mistake or wrong doing has occurred. In this manner, by taking responsibility to make things better, the child gradually develops a willingness to make things right, as he naturally begins to acknowledge his own mistakes later on.

There is always the chance in life to
re-balance the scales when a mistake or
wrong doing has occurred.

If parents on a regular basis acknowledge their own mistakes, then the child is better prepared to recognize his own mistakes. The way of taking responsibility for a mistake, besides self-correcting, is to make appropriate amends. Making amends is making things better after making a mistake. Most parents hide their mistakes from their children and rarely apologize. They assume they will lose power over their children if they acknowledge that they are not always right.

Parents can teach their children to be responsible by demonstrating responsibility. When parents are late to pick up a child, instead of explaining why they were late, they should listen, apologize, and make amends. The way to be responsible for a mistake besides self-correcting is to make appropriate amends. Making amends is simply making things better after making a mistake.

Instead of explaining why they were late,
parents should listen, apologize, and
make amends.

When a parent is late picking up a child, they can make amends by doing something extra special, such as going out of the way to get a treat. The parent might say, "I am so sorry I was late. I want to make it up by getting you a treat. Let's go get a smoothie." To make up for their mistake, parents could even offer to do one of the child's chores that day or create a fun activity. By setting an example of making amends, children are well prepared to be responsible teenagers and adults.

Children's ability to self-correct, up to the age of nine, is nurtured when they are free from the consequences of their mistakes. The ability to be responsible for mistakes and make amends is developed by experiencing again and again parents making amends for their mistakes. In this way, children not only learn to be accountable, but also learn to make it up or pay for their mistakes in a responsible and appropriate manner.

Parents can teach their children to be responsible by demonstrating responsibility.

When a teenager is late and keeps his parents waiting, besides learning to be more respectful of time, he needs to make amends. The message is simple, "If you keep me waiting, what then can you do to make my life easier?"

If a child is raised with the parent making amends for her mistakes, then automatically a teenager will be more considerate and happy to make amends. She will sometimes automatically say, "I am sorry—how can I make this up to you?"

At other times, she will immediately make a suggestion. "I am sorry for keeping you waiting. Would you like me to wash your car to make it up?"

Another healthy response is, "I am really sorry to keep you waiting. I owe you one." This means parents can think of something extra they would like over the next few weeks and the child will be happy to do it for them.

In the meantime, if the parent happens to be late, the parent might say, "I'm sorry to be late. You owed me one from last week, so now we are even."

If your teenager was not raised with positive-parenting skills, then you will have to ask, "How would you like to make this up to me?" Or, simply let him know that he inconvenienced you, and so he "owes you one."

He will catch on very soon and gladly trade the notion of getting punishments to making amends whenever appropriate. Teenagers as well do not need shame and punishment. The safety to make mistakes will pave the way for them to be more responsible no matter what age they begin.

DON'T PUNISH, MAKE ADJUSTMENTS

Positive-parenting skills do not use *any* punishments to motivate cooperation, but sometimes a parent may have to make adjustments regarding certain freedoms given to a child or teenager. If your eight-year-old son keeps jumping on couches in the living room, then you may have to restrict his ability to play in the living room unless you are around. Before making this adjustment, the five skills of positive parenting need to first be employed.

In making the adjustment, let him know that he can earn back the opportunity to play in the living room. Say something like this: "When you can cooperate by not jumping on the couches when I am with you in the room, then I will reconsider, and you can play in the living room when I am not here." In this example, the parent is not punishing; she is just adjusting the rules or guidelines.

If you have given your sixteen year old a weekend curfew of 1 A.M., and she is consistently late, then an adjustment may be required. The curfew needs to be earlier, not as punishment but because you realize your child is not yet responsible enough to stay out that late. If she is not responsible enough to remember the time and respect your curfew, then she is not responsible enough to stay out until 1 A.M.

When a teen is consistently late, adjust the curfew to an earlier time.

Long before making an adjustment, the teen should be forgiven, but also required to make amends for inconveniencing you. After using each of the five skills of positive parenting, if the teenager continues to be late, then a parent needs to acknowledge that he made a mistake in giving such a late curfew and adjust it to an earlier time.

The parent could say something like this, "I know you understand this is the curfew, and I understand that things happen, and you just forget the time. We have talked about this several times. I know you do your best, and I think you have too much independence. I am resetting your curfew to midnight. If you can respect that three times in a row, then we will move it to 12:30, and when I can trust you to remember that time, we will reconsider a 1 A.M. curfew. On the weekends I want you home by midnight."

HOW TO REACT WHEN CHILDREN MAKE A MISTAKE

When a child up to the age of nine years old makes a mistake, we should respond as if making mistakes is normal and happens all the time. The child should not be required to apologize or make amends until age ten. For example, when a child breaks something, ultimately it is the parents' responsibility because the child was not getting the supervision required. The child is not required to apologize or make amends in such a situation.

When one child hits another, the child is not to be punished nor forced to apologize. Instead, the parent should redirect the children. If there is any blame, it is always the parents' responsibility because in some way the child wasn't getting what was needed from the parent. It may be that the child didn't have enough understanding, supervision, structure, or rhythm.

When two children are fighting, instead of making them apologize and make up, just have them make up. Say something like this, "Okay, let's make up and be friends. I'm so sorry that you got hit. Let's play this game . . ."

Children demand that the other child be punished only if they themselves have been punished for their mistakes. They demand apologies only when they have been expected to apologize. When parents take responsibility instead of blaming their children, then siblings don't blame each other so much nor do they demand apologies.

It is hard to be so accepting of mistakes because most parents are not accepting of their own mistakes. They feel the need to punish, because they were punished as children, and they believe that is what happens when someone makes a mistake. Fortunately, positive parenting still works even when parents make mistakes. If a parent gets upset with a child for making a mistake, he or she can always come back and simply apologize for getting so upset and give reassurance that it is okay to make mistakes.

You could say, "Mommy made a mistake by getting so upset with you for breaking the vase. I should not have yelled at you. I am sorry. It wasn't such a big deal. We can always get another vase. It was just a mistake, and mistakes happen."

For an older child say, "I am sorry that I got so upset with you the other day. I should not have gotten that upset. A lot of other things were bothering me. It wasn't such a big deal. We can always get another vase."

If a child knocks over a vase, it's natural to feel upset about the broken vase, but a parent should be careful not to be upset with the child or with him- or herself. Some parents don't blame their children, but harshly blame themselves. Although they don't intend to make others feel bad, they unknowingly do. Parents need to model forgiveness for their children's mistakes as well as forgiveness for their own.

Positive-parenting skills are new. Most parents have no idea how to react when their children make a mistake. These are some helpful insights to react appropriately to your children's mistakes. Take a few minutes to reflect on how you would respond in the following circumstances.

If you had a great day, were well rested, and felt you had a bright future, how would you react when your child knocked over the vase?

If your child was helpful, cooperative, and always listened to you, how would you react when your child knocked over the vase?

If your child was trying to clean the vase for you and a loud alarm went off, how would you react when he or she dropped and broke the vase?

If the president of your company were to accidentally knock over the vase, how would you react?

If you allowed five boys to play football in the living room and they accidentally knocked over the vase, how would you react?

If the vase was very cheap or you were thinking of getting a new one anyway, how would you react?

If you had a blind guest who accidentally broke the vase, how would you react?

In each of these examples, your reaction to the broken vase would probably be nonshaming and forgiving. You might feel a little upset about the loss of the vase or the inconvenience of cleaning it up, but you would not dwell on it. You clearly would not be upset with your child, yourself, your boss, or your guest. You recognize that these things happen. You would naturally care more about your child's feelings, your guest, or your boss than the vase. You certainly wouldn't want anyone to feel bad. This positive, forgiving reaction is the appropriate response even when the conditions are different.

Try putting this new insight into practice by reflecting on different upsetting mistakes your child has made. Take a particular mistake and insert it into each of the seven conditions listed above and explore your reaction.

For example, if your child recently made a mess and didn't clean up after herself, ask yourself how you would react if you had a great day, were well-rested, and felt you had a bright future and your child made a mess and didn't clean up after herself. In this way, go through each of the seven conditions above to expand your awareness of how to react from your most loving self. When your children make a mistake, regardless of the circumstances, they deserve a forgiving response.

When your child breaks a vase,
regardless of the circumstances, she
deserves a forgiving response.

Now, let's change the circumstances and explore how you *don't* want to react when your child breaks a vase. Take a few minutes to reflect on how you would respond in the following circumstances, recognizing that this is *not* the way you want to respond.

If you had a terrible day, were exhausted and overwhelmed with too much to do and not enough time, and your future looked dim, how would you react when your child knocked over the vase?

If your child was always breaking things and never listened to you, how would you react when your child knocked over the vase?

If you asked your child not to play in the living room or touch the vase and she did anyway, how would you react to your child's knocking over the vase?

If a paid housekeeper were to knock over the vase after accidentally breaking several other things, how would you react?

If you asked your child specifically not to play in the living room or touch the vase, and it was knocked over, how would you react?

If the vase were extremely expensive or very special to you, how would you react when your child broke it?

If you asked your spouse to put the vase away, and he or she forgot and then it was accidentally broken, how would you react to your spouse?

Unless you applied self-restraint, you would probably react in a shaming manner. If you had a bad day, you would probably unload the stress of your day on your child. This broken vase might just be the straw that breaks the camel's back. Unfortunately, children will take your overreaction as if they were completely responsible, and they will feel disproportionally guilty.

Children assume too much blame unless someone else shoulders the responsibility.

If your child regularly doesn't listen to you, then you would probably get really upset and use this as an example of what happens when he or she doesn't listen. Your upset would not be just about the vase, but about the many times your child has not listened and a projection of the many times in the future he or she will not listen. This message is confusing and ineffective to your child, as it would be to your spouse.

When children make a mistake, it is the worst time to remind them of other mistakes they have made.

If your child defied your request to leave the vase alone, then you might feel as if you should punish the child to teach the child a lesson. As we have already explored in previous chapters, punishment and shaming messages do not work any more. There are other ways of getting your children to do the right thing. When a child is deliberately defiant, a parent needs to use the five skills of positive parenting instead of punishing. Punishment will just increase defiance.

Punishment or getting upset with your child are outdated ways to communicate. The best kind of reaction to your child's mistakes is a kind of neutral or bored look. Don't put much attention on mistakes. Instead, put attention on redirecting the child and by asking him or her to do something. In this example, you might just ask the child to help you clean the broken vase up.

Don't put much attention on children's mistakes.

When a paid housekeeper breaks your vase, appropriate amends need to be made. If the problem continues, the housekeeper should be fired. He is not your child, and you are not responsible to teach him anything. With your children, your challenge is to teach responsibility by responding in a forgiving way.

If the vase was very expensive, most parents would get even more upset. Parents need to remember that a child doesn't plan to make mistakes. If something is really expensive, the parent should protect the vase, not blame the child. If a teenager were to break an expensive vase, she should make amends, but within reason. It would be very unfair to expect your teenager to reimburse you. She doesn't make as much money as you do. A more appropriate action might be cleaning up the mess and then helping you to purchase another vase, but not having to pay the cost.

If your teen invites friends over and they damage or steal your computer, then together you need to find a way for your child to make amends. If you decide that you want some financial reimbursement, it should be proportional to your salary versus your teen's. Make the reparation fair by comparing resources. If the stolen or damaged computer cost $2,000, and you earn $1,000 a week, and your teen earns $100 a week, then he should contribute $200. If he has more in his savings account, it is not fair to raid it unless you compare what he has saved versus what you have saved.

Make the reparation fair by
comparing resources.

Besides punishment or unfair reimbursement, a parent can make the mistake of assuming that their child knows better. If you asked your spouse to put away the vase and he forgot, you might get upset, because you feel he was warned and didn't follow through. You would feel that he had been warned and knew better. Parents tend to get upset when children forget to do things. They mistakenly assume their child knew better. They forget that it is perfectly normal for kids to forget. Some children need to hear many times before they remember, and, if they are stressed, they may forget again.

DOING YOUR BEST IS GOOD ENOUGH

Children should get the message that their best is good enough and that mistakes are a natural part of the process of learning and growing. By making mistakes, we learn what is right or best for us. We should just do our best, and the rest is a process of trial and error. This is a healthy message, but it can easily be misused to shame our children as well.

Parents will often agree with the notion that doing your best is good enough. When their children fail or make mistakes, parents incorrectly assume that they are not doing their best. The child then concludes that his best is just not good enough. When criticized for making mistakes or not doing their best, children begin to feel bad about themselves.

Parents often mistakenly conclude that since their teens are more mature they should have better memories. When teens forget to do things, parents mistakenly assume that they are just not trying hard enough. Trying has nothing to do with memory. Either you remember or you don't. Forgetting needs to be treated like any other mistake.

One of the best ways of dealing with a child or teenager, who often forgets, is to ask as if you are asking for the first time. As you continue to ask in this way, the child or teen recognizes that he or she forgets. When a teen recognizes on his or her own, their ability to remember is strengthened. When you stop reminding your children what they have forgotten and they start to recall on their own, they begin to remember more of what you have asked.

For most children and teens, the more stressed they feel, the more they forget. Nagging or getting upset with them is not effective. It just creates more stress, which blocks better memory. Rewarding is a better way to increase memory. If your child or teen regularly forgets something, then at the end of each week give a reward for what they remember.

When children fail or make mistakes, parents
incorrectly assume that they are not doing
their best.

When children make mistakes, disappointed parents unknowingly give a variety of shaming messages that make children feel bad, wrong, inadequate, or unworthy. These are some common messages:

You know better than that.

You can do better than that.

You should have known better than that.

How could you have forgotten?

I've told before.

I warned you.

If you would only listen to me . . .

What's wrong with you?

I've seen you do better.

What's the matter with you?

You just don't listen.

Parents incorrectly assume that when their children are making mistakes, not behaving or performing up to their expectations, they are not doing their best. Telling children that you think they are not doing their best is shaming. It is using disapproval as a way to motivate children. As we have explored, using guilt to manipulate children is not only unnecessary, but, for children today, it just doesn't work.

Positive parenting recognizes that children are always doing their best. When they make mistakes, that is part of their learning process. When they are out of control, they are not getting what they need to be in control. Regardless of what children are doing wrong, at that moment, they are doing their best.

Positive parenting recognizes that children are always doing their best.

Children don't wake up thinking, "How can I do my worst today? What can I do to really fail? How can I disrupt my parents' life and get them to hate me?" No one thinks this way unless they are deeply hurt and wounded. Even then, a child would be doing his or her best, albeit in a misguided manner, to get his or her needs met.

To do one's best does not mean that one's full ability has been expressed. It just means that based on your resources at that time, you did your best. Let's take a simple example to make this point clear.

Yesterday, I did my best writing. I wrote thirty pages in one day. For many writers, this is really great. When I started writing this book, I did my best, but put out about three pages a day. Today, I was tired, but I did my best and wrote about five pages. One day I wrote three pages. Later in one day, I wrote thirty pages. Then the next day I wrote five pages. Yet, I was doing my best every day.

Considering this example, you will see that the result or outcome is not an accurate measure of doing my best. In a similar manner, it is a mistake to measure our children's best by their performance. The secret for making it okay for our children to make mistakes is to recognize that they are always doing their best.

WHEN IT IS NOT OKAY TO MAKE MISTAKES

When making mistakes is not acceptable, children react in a variety of unhealthy ways. The following list contains four common ways our children react, when their mistakes are not accepted:

1. Hiding mistakes and not telling the truth.

2. Not setting high standards or taking risks.

3. Defending themselves by justifying mistakes or blaming others.

4. Low self-esteem and self-punishment.

These four reactions can be avoided when children get a clear message that it is okay to make mistakes. Children come into this world with the ability to love their parents, but they cannot love or forgive themselves. Children learn to love themselves by the way they are treated by parents and by the way parents react to their mistakes. When children are not shamed or punished for their mistakes, they learn that they don't have to be perfect to be loved. They gradually learn the most important skill: the ability to love themselves and accept their imperfections.

Children learn to love themselves by the way
they are treated and by the way parents react
to their mistakes.

When parents apologize for their own mistakes, children automatically forgive. Children are preprogrammed to forgive their parents, but not themselves. If parents don't make mistakes and then apologize, children will never learn how to forgive. If parents make mistakes and don't apologize, children will blame themselves. Without enough practice forgiving their parents, children do not learn how to forgive themselves.

Self-forgiveness dispels the darkness of guilt. It is learned by having the opportunity to experience repeatedly that their parents make mistakes and are still lovable. When children see their parents as imperfect but still lovable, then at the age of ten when they suddenly become more self-aware, and thus aware of their own imperfections, children will not be so hard on themselves.

At about nine years old, children start feeling embarrassed when their parents do things they think are strange, for example, when Mother sings in the grocery store. As they become more self aware, they are suddenly more aware of what other people think. When children have been raised with a forgiving attitude, they are more forgiving of their own imperfections. Let's explore in greater detail the four ways children react when it's not okay to make mistakes.

HIDING MISTAKES AND NOT TELLING THE TRUTH

When children fear punishment or the loss of love in response to their mistakes, they learn to hide their mistakes. Rather than face punishment, they would rather hide what they have done and hope that they don't get caught. This leads to lying. This tendency to hide mistakes gradually develops a split inside. The child has to live in two worlds. In one world, she may be getting her parents' love, and in the other, she believes that, if her mistake were discovered, she would lose love. This has the impact of negating the love she does get.

When children do something wrong and hide their mistake, a part of them feels unworthy of their parents' love. Even when the parent does love, support, praise, or acknowledge the child, a little part of himself feels, "Yes, but you wouldn't say that if you knew what I did." This feeling of unworthiness continues to push away the love and support that is bestowed on this child. Although there is love to support him, he is unable to let it in. To the extent that children have to hide their mistakes, they will invalidate the real love and support that is available to them.

> When children hide a mistake, a part of
> themselves is unable to let love in.

Children depend on support from their parents to feel powerful and confident. When this support is cut off, the child becomes increasingly insecure. It is very wounding when anyone tells a child, "Now don't tell your mother or father. This will be our little secret." If the child doesn't feel safe to reveal all to her parents regarding her mistakes or the mistakes of others, then a wall goes up separating the child from her parents' support.

It is even more wounding when one parent demands that a child keep a secret from the other parent. It may even be a casual message, "Okay, I will give you this ice cream, but don't tell your father." This message brings the child too close to the mother and disconnects the child from the father.

It is even more wounding when the request for secrecy is backed by the threat of punishment. For example, when a child is mistreated by his father, the father may say, "I will hurt you if you tell your mother." Unless this child tells the mother, the concealment is more wounding than the mistreatment that he received. Mistakes happen and can be healed, but if the child feels he can't be open with his parents, then the healing stops.

CHILDREN OF DIVORCED PARENTS

Children of divorced parents often feel unable to share their feelings and experiences. When living in different homes, children must get the message that it is okay to talk about what happens at the other house. When it is not safe to share, a split occurs: They can only share a certain part of themselves at their mother's house, and a different part at their father's.

Children get the message that it is not safe to share when they tell a story to mom about dad, because mom gets upset or jealous and vice versa. As the child is describing what a fun time she had with dad at the fair, the mother is fuming inside that he didn't make her do homework. Children can sense these disapproving messages and just stop sharing. To make matters worse, mom calls her ex-husband and complains to him. Next time the child does something fun with her father, the father may directly ask her to not tell mom.

It is hard to make it safe for children to share the details of their lives. If parents become negative, critical, or disapproving, the children will stop sharing. At this point, not only do they become more insecure, but the parent also loses some of his or her power to assert leadership.

The more a child or teenager feels free to share everything with you without getting hurt or hurting someone else, the more the child will be willing to cooperate with you. Remember, when children feel safe to be themselves, they are automatically motivated to cooperate. To increase their power of influence over teenagers, parents must back off

250

John Gray

from giving judgments and solutions so that their children will continue to come to them and share their world.

NOT SETTING HIGH STANDARDS OR TAKING RISKS

When children get shaming messages regarding their mistakes, they are often afraid of making more mistakes. To protect themselves from the painful consequences of making a mistake, failing, or disappointing their parents, they play it safe. Instead of setting high standards that they may not achieve, they do what is predictable and secure. Living in a comfort zone, they not only sell themselves short, but also become bored because they are not being challenged.

Living in a comfort zone, children not only sell themselves short, but also become bored.

Other children react to shaming messages by being high achievers. They can't bear the pain of being less than expected or disappointing their parents, so they try harder than they need to. They may produce results, but they are never happy. What they do is never good enough for them, and they are never good enough. It is not uncommon for these children to make all As and one B in school and to come home and hear this wounding response, "Why did you make this B?"

A father might say to his high-achieving son, who missed a football pass, "If you had caught that last pass, your team would have won." Parents often ignore the positive and focus on the negative behavior. These kinds of negative messages from childhood are commonly heard in a counselor's office when helping adults to deal with anxiety and depression. In most cases, it is not that their parents didn't love them; they simply didn't know a better way of showing their love. They didn't know what they were doing to their kids. Many mistakenly thought they were motivating their children to try harder in a healthy way.

> Parents mistakenly think they are helping
> when they give shaming messages.

When children don't feel safe to make mistakes, they will tend to back off from taking natural and healthy risks. Children need to take risks to develop who they are. Without the safety to make mistakes, they hold back and often don't know why. They need a safety net. Even high achievers will feel unsafe to take risks in other areas of their lives.

Without this inner security, they may say, "I don't like parties," but under the dislike is the fear of being rejected. Rather than risking the pain of feeling inadequate, they would rather not come out and share themselves. Underlying their thinking is a fear of losing what they have if they were to make a mistake or to fail.

Fear and insecurity are not always the cause of resistance. Some children are naturally shy and take longer to form relationships or don't readily take risks. Receptive children tend to be shy and resist change. Sensitive children have a greater fear of being rejected and naturally take more time to open up to potential friends. Certainly, these tendencies to hold back are magnified when children get the message that it is not okay to make mistakes.

> Natural tendencies to hold back are magnified
> when children get unforgiving messages.

To avoid the pain of disapproval, some children will just stop caring what their parents think. This is often the case of the teenager who shares nothing with her parents. She is used to being corrected and criticized her whole life. Now that she is freer and doesn't need her parents as much, she goes to her friends to find acceptance. She rebels and makes a point of not needing her parents' approval anymore. Behind this tendency is years of having to hide herself or to hold back in order to be accepted.

Regardless of what the wounding parents have done in the past, they can make up for it at any age by using the five skills of positive parenting and applying the five positive messages. Parents need to remember that it is okay that they make mistakes, too. All parents do the best they can with the resources they have.

JUSTIFYING MISTAKES OR BLAMING OTHERS

Growing up in an unforgiving environment makes children defensive. They either actively defend by justifying their mistakes or blame someone else. When a child is asked to stop hitting his sibling, because he is afraid of getting punished, he blames the sibling. He says, "He hit me first." This defensiveness is natural, but it is magnified by punishment. When a child isn't afraid of punishment and a parent asks him to stop hitting, he will more readily listen and cooperate. He doesn't feel a great need to blame others or justify his actions.

As adults, the only way we can learn to self-correct our behavior is by taking responsibility for our mistakes. As long as we justify our mistakes by blaming someone else, we cannot self- correct. Although we are adults, we behave like children who have been raised in an unsafe environment.

As long as we justify our mistakes by blaming
someone else, we cannot self-correct.

Carol came to me for counseling. She didn't know
whether to stay with her new husband, Jack, or leave. On
several occasions, he had become angry and then violent.
The occasion that triggered her visit was when he had taken
all of her belongings and thrown them out of the house.
Later, he was remorseful and wanted her to come back. He
clearly loved her when he wasn't upset, but he probably
wasn't ready or capable of having an adult relationship.

She wanted to know what I thought. I told her I would
need to talk with him. When they both came in for a session,
I asked him if he would ever do this again. He was very definite in his response. He said, "What I did was wrong, but
what she did was wrong, too. As long as she doesn't say the
things she said, then I will never become violent again."

After much discussion, he would not budge. I tried to help
him see that what he did was wrong no matter what she did to
provoke it. Jack could not accept that, and, as a result, Carol
was able to see clearly that he was too immature to be in a
marriage. In Jack's mind, his violent and abusive behavior was
justified by Carol's rejecting comments. As long as her behavior justified his, he could not truly self-correct. Clearly, as a
child, Jack didn't grow up in a forgiving environment. He
never learned to be responsible and to self-correct; instead, he
learned to defend himself by blaming others.

When children don't feel safe making mistakes, too
much time, energy, and conversation is wasted on defending
what happened, explaining why it happened, and what
should happen because it happened. All this misery for both
children and parents can be avoided by making it safe to

make mistakes. When it is okay to make mistakes, instead of defending, children are open to listening to what parents want him to do. Looking back and trying to teach children what they did wrong is a dead-end street—it goes nowhere.

When it is okay to make mistakes, instead of defending, children are open to listening.

When we justify our mistakes and blame others for our problems, we reinforce the mistaken notion that we are powerless to solve our problems. When we make others responsible for our problems, we forfeit our power to heal our wounds, learn from mistakes, and proceed in our lives to get what we want.

TEENS AT RISK

I was once asked to develop a training program in Los Angeles for teens at risk who had been molested or abused in some serious manner. All these children were experiencing different kinds of behavior problems and low self-esteem as a result of the mistreatment they had suffered. To teach the first weeklong workshop, I insisted that a mix of children be present. For my purposes, it was healthy to mix teens from dysfunctional families with teens from more loving and supportive families.

Nobody was singled out as coming from a dysfunctional family or having been abused or molested. Gradually, as teens were able to talk about the negative messages they received growing up, they discovered that their pain was the same. Without pointing out the really "abusive experiences" that some of the teens had experienced, the teens realized that 90 percent of their issues were the same. They all felt misunderstood, unimportant, unappreciated, and unfairly treated at times.

Quite often, well-meaning therapists and parents put too much emphasis on one negative, abusive experience and a client concludes that all of his or her problems come from that experience. To blame one's unhappiness on a single circumstance is misleading. This illusion was dispelled by being with more fortunate teens, who were gradually willing to disclose their inner feelings of fear, anger, disappointment, pain, hurt, rejection, injustice, guilt, sorrow, loss, resentment, and confusion. Although the teens at risk had certain unfortunate circumstances, they discovered that much of their pain was the result of common misguided parenting approaches.

Since teenagers are unaware that others feel the way they feel, they continue to blame their negative feelings on their past. Many adults today continue to blame their present pain or misfortune on their past. This blaming mentality blocks them from finding the power to change their lives.

Workshops, counseling, and support groups can help to heal the pains of the past, but, to minimize the wounding in the first place, parents can strive to create a safe environment for their children to make mistakes. Once a child has developed the tendency to defend by blaming, a parent can help by applying the skills of positive parenting and by becoming a role model of taking responsibility for past mistakes. When parents become more responsible for mistakes and less blaming of others, children automatically become less defensive and blaming.

LOW SELF-ESTEEM AND SELF-PUNISHMENT

Children learn to regard themselves by the way they are treated. Neglecting children clearly wounds their self-esteem. When children don't get what they need, they automatically begin to feel unworthy. Even when they are not neglected, they may begin to feel unworthy and inadequate. When concerned parents feel frustrated, angry, hurt, embarrassed, or worried by their children's behavior or mistakes, then their children feel unworthy of love or inadequate in some way.

Instead of feeling healthy self-esteem, these children feel they don't measure up. In trying to measure up, they desperately attempt to be perfect to please their parents. They can never succeed, because no one is perfect. They may be very good in their behavior, but it is at the cost of their self-esteem. Unable to please their parents, they feel less than or inadequate.

Children may be very good in their behavior,
but at the cost of their self-esteem.

Negative emotions arise in parents when children don't measure up to their expectations. No matter how much parents say they love a child, when they get upset with a child's mistakes or deficiencies, the child gets the message. The only way children can measure themselves is through their parents' reactions. If children are to feel good about themselves, parents must constantly adjust their expectations so that negative emotions don't continue to surface.

Children feel okay when parents feel
everything is okay.

When parents are happy, accepting, respectful, understanding, caring, and trusting children get the clear message that they are good enough. Naturally, they feel good about themselves. They feel safe to experiment and be all that they can be. They trust themselves and have greater confidence. They are more relaxed, because they do not always feel they have to measure up to some impossible standard. They simply and innocently feel that they are who they are supposed to be, and they are doing what they are supposed to be doing. The freedom to make mistakes without consequence in the first nine years of life generates a comforting feeling of security.

Imagine right now how you would feel if you could do anything without getting into trouble or that whatever you attempted would be good enough. How would your life be different if you had no fear to hold you back or guilt to hold you down? Feel the freedom and peace you would have to be yourself, and the joy and confidence you would experience in doing new things.

This is the gift you can give to your children during the age of innocence. This feeling, when allowed to flourish for nine years, never goes away. Even though children grow up and learn to be responsible for mistakes, the feelings of innocence remain as a foundation. As adults, when they make mistakes, they easily come back to self-forgiveness and self-correction. They have a greater sense of compassion and respect for others, because they are not concerned with defending themselves.

When innocence is experienced for nine
years, the feeling never goes away.

When parents take responsibility for a young child's mistakes, that child learns that he is innocent. On the other hand, when he is punished or shamed for mistakes, he begins to feel unworthy of love and inadequate. If he has been punished for mistakes, then he gradually learns that to be worthy of love after making a mistake he must be punished.

Many adults hold back from taking risks, because they are so hard on themselves when they make mistakes. They suffer anxiety because they are so afraid of the misery they'll experience after making a mistake. They often have a nonspecific fear that looms over them whenever they are faced with the possibility of making a mistake. These adults were often punished as children for mistakes, and continue to feel the fear of punishment as a result. Even though their parents are no longer around, they still feel the fear. When they do make mistakes, they are generally much more demanding of themselves than others would be.

When children are punished for mistakes, they continue to feel the fear throughout their lives.

In some cases, to avoid being so hard on themselves, they are hard on others. To protect themselves from punishment, they blame and punish others. This tendency can also go in the opposite direction. A person can be very accepting of mistreatment by others, because she feels so unworthy of loving and good treatment. Others may mistreat her, but she feels that in some way she deserves it. She may be very forgiving of others, because she has low self-esteem and feels worthy of punishment.

Whatever punishment goes in eventually comes out,

either on others or on themselves. Girls particularly punish themselves, while boys feel more justified in mistreating or punishing others. A girl might punish herself by getting involved with someone who hurts her, or she may just beat herself up with negative thoughts and self-criticism after making a mistake. A boy has a greater tendency to blame others for his mistakes and punish outwardly. Any child, boy or girl, could demonstrate either of these tendencies. Whether boy or girl, when a child is punished for making mistakes, the outcome is that the child is unable to forgive him- or herself and others for those mistakes.

MAKING IT OKAY TO MAKE MISTAKES

Without an understanding of the five skills of positive parenting and the importance of making it okay to make mistakes, shame and punishment have been the only tools parents had to control and protect their children. Parents thought that too much praise weakened self-esteem and made children egotistical. They believed that if they did not punish their children for their mistakes, then their children could not learn right from wrong. While these notions are clearly outdated and abusive today, there was a time when they were the only tools that would work.

In putting the new ideas of positive parenting into practice, it is important to remember that parents make mistakes, too. Our children are incredibly adaptable; they have within themselves the power to adapt and be all they can be regardless of a parent's mistakes. Life is a process of making mistakes and encountering the mistakes of others. It is through this very process that our children can become all that they can be.

If you recognize ways that you have wounded your children, instead of feeling guilty, forgive yourself, as you would have your children forgive themselves for their mistakes. Remember that you are always doing your best with the resources that you have. Be careful not to beat yourself up with this information. Instead, be happy that you now have a new and better approach than what was given to you.

Instead of wasting energy blaming your parents for their mistakes, forgive your parents, as you would want your kids to forgive you, and use that energy to continue your research into being a better parent. Use *Children Are from Heaven* as a resource you can return to again and again. Take parenting workshops and create a parents' support group to work with other parents using this positive-parenting approach. As you take time to move through your learning curve, you will automatically become more accepting of your child's unique learning curve as well.

11

It's Okay to Express Negative Emotions

All children experience negative emotions in reaction to life's challenges and restrictions. Negative emotions are a natural and important part of child development. They assist children in making necessary adjustments in their expectations to help them accept life's limits. Positive parenting skills, such as listening with empathy and time outs, provide opportunities for children eventually to learn appropriate ways to express their negative emotions.

Past parenting approaches attempted to control children by suppressing feelings. Shaming or punishing children for being upset subdues their passions and breaks their will. Giving children permission to feel with the positive message that "it is okay to express negative emotions" empowers children. It awakens and strengthens willpower and gives a sense of direction. Unless a parent knows the skills of positive parenting to create cooperation, this extra power can be counterproductive.

Shaming or punishing children for being upset
subdues their passions and breaks their will.

When children's feelings are honored and heard, a stronger sense of self can develop, but a false sense of power can also be created. If a child throws a tantrum and gets her way, then the permission to express negative feelings not only spoils the child, but creates an inner insecurity.

Parents must be careful not to placate a child to avoid having to deal with a tantrum. Giving a child permission to express negative emotions must be balanced by strong parents who are not threatened by such tantrums. When a parent learns how to handle a child's tantrums with the five skills of positive parenting, the permission to express feelings is then a tremendous gift.

**Parents must be careful not to placate a child
to avoid having to deal with a tantrum.**

When their children throw tantrums, most parents mistakenly conclude that their children are bad or that they are not good parents. Learning to express, feel, and release negative emotions is an essential skill every child needs to learn. Learning to manage negative feelings in this way awakens inner creative potential and prepares children to cope successfully with life's challenges.

THE IMPORTANCE OF MANAGING FEELINGS

The most important element of learning to manage negative emotions is to make them acceptable. Although negative emotions are not always convenient or pleasant, they are a part of growing up. By first learning to express, feel, and release negative emotions, children eventually gain an inner awareness of their feelings and can more easily feel and release negative emotions without having to act them out in any manner.

The most important element of learning to manage negative emotions is first to make them acceptable.

By learning to feel and communicate negative emotions, children most effectively learn to individuate from their parents (they develop a strong sense of self) and gradually discover within themselves a wealth of inner creativity, intuition, love, direction, confidence, joy, compassion, conscience, and the ability to self-correct after making a mistake.

All the skills of living that make a person shine in this world and achieve great success and fulfillment come from staying in touch with feelings and being able to let go of negative feelings. Successful people fully feel their losses, but they bounce back because they have the ability to let go of negative feelings. They are able to manage their negative feelings without having to suppress them or get lost in them.

Most people who do not achieve personal success are either numb to their inner feelings, make decisions based on negative feelings, or just remain stuck in negative feelings and attitudes. In each case, they are held back from making their dreams come true. To stay in touch with your inner passion and your power to get what you want and need in life, it is essential fully to feel. Positive parenting techniques gradually teach your children to manage their inner negative emotions and to create positive emotion.

Unsuccessful people are either numb to their
inner feelings or remain stuck in negative
feelings and attitudes.

Passion means intense feeling. Passion can be sustained in life if we are able to manage our negative emotions successfully. If we learn to suppress our negative emotions, we gradually lose touch with our ability to feel positive emotions as well. We lose our ability to feel love, joy, confidence, and inner peace.

Failure is inevitable in life when adults make decisions or take action based on negative emotions. A successful adult needs to learn how to feel negative emotions and then release them. As a result, positive feelings return, and they are then ready to make healthy and more successful decisions.

LEARNING TO MANAGE FEELINGS

Negative emotions are always okay, but they must be expressed at the appropriate time and place. It is not acceptable for a child to dominate the family with demanding emotional tantrums. Parents must be strong, but at the same time create opportunities for children to have the tantrums they need to have.

Young children need to be expressive and to communicate their negative emotions. By being empathetic and giving time outs, wise parents provide their children with regular opportunities to feel and express their negative emotions fully. Although it is not okay to act out from negative emotions, it is the parents' responsibility to control the child's behavior while also making it safe for the child to express negative feelings. When parents use the positive parenting skill of listening and time out, children gradually learn to regulate and manage when, how, and where their emotions are expressed and communicated.

Remember God makes children little so you
can pick them up and put them in a time out.

Although it is fine to express negative emotions, it is not acceptable to act them out or express them anywhere and in any situation. It is appropriate to express negative emotions when a parent can and is willing to listen or during a time out. Gradually children learn to regulate their need to express at times when parents can listen.

With regular time outs from two to nine years old, children will gradually learn how to regulate when and how to communicate negative feelings. Though this may seem like a long wait, it is not. Most adults who come for counseling (and many more who don't) have still not learned how to manage emotions successfully.

To regulate children's tantrums, a parent needs to give regular time outs. If a parent doesn't give enough time outs, children inevitably will act out at times when it is not easy or possible to give a time out. With this awareness of the importance of expressing emotions to release them, parents are more willing to listen to their child's negative emotions and give regular time outs. They clearly recognize that children need to throw tantrums in the appropriate setting of a time out. They no longer feel the need to placate them to avoid confrontation and tantrums.

When positive parenting skills are used, children learn that it is okay to express negative emotions, but mom and dad are still the bosses. When the parent determines that it is time to finish a negotiation, it is time to stop expressing. If a child can't stop expressing, then a time out will enable a child to release her feelings behind a door. As we have already explored in Chapter Eight, within a few minutes, a child will express and feel the necessary emotions of anger, sadness, and fear and automatically come back to feeling more in control and cooperative.

COPING WITH LOSS

Children tend to have more intense emotions than adults do, because they do not develop the ability to reason until they are nine years old. They cannot reason away their emotions. If someone is mean to them, they feel temporarily that everyone will always be mean to them, or that they somehow deserve to be treated that way, and they will always be treated that way.

They don't have the reasoning capacity to realize that one person being mean doesn't mean everyone will be mean. They can't reason that if someone is mean to them it may have nothing to do with them at all. It could be that person is just having a bad day. Since they don't have the ability to reason, their feelings of loss are much greater.

Often parents unknowingly wound their children by minimizing their feelings of loss. One of the best ways for parents to empathize is simply to accept that when children are upset, there are valid reasons for their upsets from their perspective. Trying to talk them out of their feelings is not necessary. By simply allowing children to express their emotions, they can feel better and then be receptive to reasonable reassurance.

When children are upset, there are always valid reasons from their perspective.

Most adults today understand that to cope with a great loss, feelings of anger, sadness, fear, and sorrow are not only natural, but lead us to feeling better. When we don't get what we want or we lose someone or something special, sometimes we just need to have a good cry. Feeling and then releasing negative emotions helps us to accept life's limitations. Likewise, to learn to accept the limits their parents impose, children need to have a good tantrum. Regular tantrums are the way young children express and thereby feel their negative emotions. Eventually, they learn to feel inside without expressing or acting out their negative feelings.

Regular tantrums are normal and natural up to the age of nine. If children do not have the opportunity to throw enough tantrums, instead of outgrowing this phase of development, they continue to throw tantrums for the rest of their lives. Children today are more sensitive than ever and have an even greater need to express feelings. So many of the new problems we witness in children from hyperactivity and violence to low self-esteem and suicide will be solved as children get this support and learn to manage their emotions successfully.

WHY EXPRESSING EMOTION HELPS

The act of expressing negative emotions enables children to feel. Children become aware of their feelings by first expressing negative emotions. Feeling is the ability to know what is going on inside ourselves. Getting in touch with feelings makes us more aware of who we are, what we need, wish, and want. The ability to feel helps us to recognize and respect what others need, wish, and want as well. Listening to our children express negative emotions helps them to develop their ability to feel.

> Getting in touch with feelings makes us more
> aware of who we are, what we need, wish,
> and want.

Creating safe opportunities for children to express and feel emotions of anger, sadness, and fear reconnects our children to their basic inner need for their parents' love. Suddenly, getting their parents' love becomes much more important than what they were upset about. When a child throws a huge tantrum because he or she can't have a cookie, this child has just temporarily forgotten who is the boss and the importance of love over getting a cookie. Supporting children in expressing negative emotions will always bring them back to feeling their need for their parents' love and a strong desire to cooperate and please their parents.

> When children throw tantrums, they have
> temporarily forgotten who is the boss and
> the importance of being loved.

With an increased awareness of their need for love, suddenly the need for a cookie diminishes, the tantrum dissipates, and the child becomes more cooperative. In this way, children become once again grounded in their true self, which is happy, loving, confident, and peaceful. They are aware once again of their need for their parents' love and their innate willingness to cooperate and please them. All this comes from creating a safe opportunity for children not to get their way, to throw a tantrum, and not to risk punishment or the loss of love in the process.

Listening with empathy and giving time outs are the most

powerful ways a parent can give the message that it is okay to express negative feelings. Even when children resist going into a time out, it is okay. They may get angry and say mean things. That is okay. A time out is an opportunity for children to resist with all their might and then finally surrender to the parents' control. It is important for children to know that they are not bad for resisting a time out or for having to take one. It is seen simply a natural part of growing up.

You must make sure that you are not placating a child to avoid a tantrum, otherwise tantrums will happen when you don't have an opportunity to give your child a time out to deal with his or her feelings. Children need to feel that they are in their parents' control. When children stop feeling in their parents' control or sense that their parents can't control them, they seek control by becoming demanding or by throwing a tantrum.

THE POWER OF EMPATHY

To assist our children in expressing their negative emotions, parents must learn to develop empathy. It is not enough to love our children; we must also be able to communicate that love in meaningful ways. Although love is most important, how we demonstrate our love makes the difference. Communicating empathy is one of the greatest gifts a parent can give.

Empathy draws out children's negative emotions and actually heals them. Empathy communicates the message that your feelings are valid. Parents are always in a hurry to reassure their children that things are all right. Before children can let that message in, they must first feel heard. Children must experience just for a few brief seconds that you understand their perspective, and then they can absorb your reassuring perspective.

> Empathy is that magic switch that opens
> children up to receiving reassurance and
> guidance.

When a child is upset because he didn't get what he wanted, many parents are too quick to make him feel better right away. This approach prevents the child from staying in touch with his feelings of loss. When a child feels his loss, he is most receptive to drawing in the empathy he needs. Remember, it may only take a few seconds. After he receives the empathy, his emotions change. He either moves to a deeper level of negative emotions (from anger to sadness to fear), or he feels better. Not only does the child get what he needs to feel better, but he also experiences that he has the power to let go of negative emotions to feel better.

> Giving empathy sometimes only requires a
> few extra seconds of silent caring and
> understanding.

When parents are always quick to give solutions so that a child feels better, the child misses the opportunity to learn how to let go of negative emotions and to find the positive feelings deep inside. When parents give solutions, children become dependent on solutions to feel better and don't learn to accept life's setbacks with a positive attitude. Our children become too dependent on getting what they want to be happy, rather than learning to be happy when there is love, no matter what the outer circumstances are.

When parents give empathy rather than solutions, children develop the ability to adjust to any negative circumstance or disappointment without having to fix things. By giving children empathy before helping them solve their problem, children develop the ability to let go of negative emotions and to feel better and then to move on to solving the problem. Most adults today still have not developed this ability, because as children they did not get the empathy they needed. When children only get solutions to their problems, they stop going to their parents for help.

What our children need most is silent understanding, caring, and a little expressed validation.

THE FIVE-SECOND PAUSE

Sometimes all it takes is just to slow down and not try to solve your children's problems. When they are angry, sad, disappointed, or worried, instead of telling them how to feel better about the situation, the positive-parenting approach is to do nothing but simply feel for five seconds what they are probably feeling. When the child feels disappointed, instead of trying to cheer her up it is better to just let her feel disappointed and feel what she is feeling. Pause for five seconds and simply feel what you think, feel, or sense she is feeling.

Instead of giving a solution, just feel her disappointment along with her and then after five seconds say something simple like, "I know, it is really disappointing." This gives the child a clear message that disappointment is a part of life. Nothing dramatically wrong has happened. Very quickly a child's mood will begin to change. More than anything, children need a clear message that for life to be okay, they don't need positive things to happen all the time.

When you have a quick solution, it not only minimizes their feelings, but also reinforces a sense of incompetence. If it is so easy for you to solve what is wrong with them, either they feel wrong for being upset, or they feel inadequate for not dealing with the situation the way you suggested in your quick fix. Certainly, a solution is given with love, but, if administered right away, it can have the opposite effect you are intending. Solutions are fine once the child has started to feel better or begins to ask for another way of dealing with the situation.

When your children stop listening to you, it is clearly because you have been giving too much advice.

Sometimes even when a child is asking for a solution, he is still not ready to receive one. Although he is asking for a solution, he really needs more empathy. The child might say in an upset tone, "I don't know what to do." At this point, most fathers will jump in and offer a solution. Then, quite often, a father will get into a power struggle or argument. His child will inevitably respond to the "solution" with the resistance of "but." The "but" meaning, "But that doesn't apply," or "But that doesn't work because . . . ," or "But you just don't understand."

When a child says, "You don't understand," this will generally put most fathers on the defensive and, instead of being there for the child, the child will eventually have to be there for the father. The father begins to demand that the child understand why he is right. This is not the right arrangement. The parent is responsible for the child and not the other way around.

When a child says, "You don't understand," stop immediately. Bite the bullet. Agree with the child. He is right. At that moment, you are not understanding or feeling what he is feeling. Instead of retracing your steps and explaining that you do understand what he has said or his situation, just stop and agree. Say something like this, "You're right, I don't understand. Tell me again." This time, back off from giving solutions and focus on giving empathy.

These are a few examples of the common solutions mom or dad might give and alternative statements to communicate empathy.

Giving a Solution	*Empathizing*
Don't cry.	Pause five seconds then say: I know . . . it's disappointing.
Don't worry.	Pause five seconds, then say: It is difficult. I know you are worried.
It will be okay tomorrow.	Pause five seconds, then say: It is hard. I know you're disappointed.
It's not such a big deal.	Pause five seconds, then say: I know you feel hurt. Let me give you a hug.
Hey, you can't win them all.	Pause five seconds, then say: I know you're sad. I would be sad, too.
Hey, come on . . . that's life.	Pause five seconds, then say: You have a right to be angry. I would be angry, too.
It could be a lot worse.	Pause five seconds, then say: I can see you're afraid. I would be afraid, too.

You'll do fine . . . everything will be all right.	Pause five seconds, then say: I know you are scared. It's scary.
It's not that important anyway.	Pause five seconds, then say: It's okay to be jealous. I would feel jealous, too.
You'll get another chance.	Pause five seconds, then say: If that happened to me, I would be disappointed, too.

WHEN CHILDREN RESIST EMPATHY

When giving empathy messages, a child may resist and correct the parent by saying, "That is not what I am feeling" and then proceed to talk about how she is feeling. Even if you believe you were right, it is important not to interrupt her flow to defend your observation. The point is not to be right, but to help your child express her feelings.

When a child is upset, there are always a variety of feelings inside. By pointing out one feeling a child may be having, he will quickly move on to another by saying, "No I'm not angry; I am sad." Although it may feel as if he is not listening to you, try to remember that this is the time when he needs you to listen to him. When a child's feelings shift and move on, it is a good thing.

When children resist, try to remember that
this is the time when they need you to listen
to them.

With a greater awareness of their different feelings, children can more quickly let go of them. If a child rejects your empathy or corrects your statement with a comment like,

"That's not how I feel . . . ," make sure you don't get into an argument. Just accept the resistance and keep listening. As the child continues to talk about his or her feelings, you have succeeded in helping.

If a child has not had ample opportunities to express feelings in the recent past, then empathizing with current feelings may open up a whole box of feelings. All the things that have bothered the child over the last several months or even years may begin to come up. This is good. Let it happen.

Just listen. Once it is out, the child will soon feel better. Parents often make the mistake of trying to cut children off by pointing out that they are not correct or they are getting off the subject. That does not need to be said. Instead, just let them talk, and eventually they will be able to forget the past and appreciate you for listening.

WHEN PARENTS EXPRESS NEGATIVE EMOTIONS

Negative emotions tend to stimulate emotional reactions in others. When someone is sad, we feel sad. When someone is angry with us, we often feel anger in return. If someone is really scared, we may suddenly become unsettled or worried. If we are already upset, then another person's emotions will stimulate our own feelings.

This helps explain why it can be so difficult to listen to a child cry or why being empathetic is easy sometimes but difficult at other times. If you have had a bad day and feel unsettled emotionally, listening to your children may become more difficult. All it takes is for your child to get upset, and suddenly your unresolved feelings come up in reaction to your child.

For example, if you are already frustrated because there seems to be too much to do and not enough time, you will have a greater tendency to overreact to your children. If your child is upset and frustrated with her homework and you are trying to help her, it will trigger your own frustrated feelings. Suddenly you'll find yourself lashing out in a frustrated manner with your child. What started out as a loving gesture of help turns into a painful argument.

When parents react to their children's negative emotions with more negative emotion, it doesn't make children feel safe to express negative emotion. When parents express negative emotion, they are bigger, louder, and more powerful. Strong adult emotions intimidate children into not expressing negative emotions. Eventually, children become numb to their feelings when it is not safe to express them.

> When they express negative emotions,
> parents are bigger, louder, and more
> powerful and intimidate children.

Some mothers regain control of their children by letting out a loud, high-pitched, emotionally charged scream. Suddenly, their children behave. They suddenly behave because it is suddenly not safe to feel negative emotions, and, the child becomes obedient out of fear. While this method works in the short run, it numbs children to their inner feelings and suppresses their inner willpower.

> Strong adult emotions make it unsafe
> for children to feel.

Fathers establish control and dominance by yelling in a mean, angry tone. Suddenly their children behave immediately. Again, the children behave because it is not safe to express negative emotions. They become temporarily obedient out of fear. Children often suppress anger, because their parents get angry in return. While this kind of intimidating control used to work, it doesn't work today. Children raised with this kind of intimidation either become rebellious later and resist cooperation or they become submissive and lack direction.

If we are to help our children manage their feelings, we must also manage our own feelings. To overcome our tendency as parents to expel our unresolved feelings on our children, we need to take time for ourselves to cope with stress and process unresolved emotional issues. Unless we take time for ourselves to cope with stress, we block our children from learning to manage their feelings.

To help our children manage their feelings, we must also manage our own feelings.

Parents cannot patiently listen to children express resistant feelings of anger, sadness, and fear if they are holding in unresolved feelings of anger, disappointment, frustration, worry, or fear. If parents are resisting dealing with feelings within themselves, then they will automatically resist dealing with their children's negative feelings.

Children cannot get the empathy they need when a parent is resisting what he or she hears. When children get the message that their emotions and needs for understanding and affection are an inconvenience, they will begin to suppress their feelings and disconnect from their true self and all the gifts that come from being authentic.

Until the parents deal with their own feelings, they are less effective in helping their children to manage feelings.

Yet, if they do take time to cope with stress and nurture their own adult needs for conversation, romance, and independence, they are able to come back and give so much more to their children. When parents take care of their own needs first, then they are most capable of putting into use the five skills of positive parenting.

THE MISTAKE OF SHARING FEELINGS

In the 1970s, adults started becoming aware of the importance of getting in touch with feelings. Just as adults needed to be in touch, we recognized that our children needed to learn about feelings as well. In an attempt to teach children how to get in touch with feelings, "enlightened" parents started sharing feelings with their children. Though the goal was correct, the process was not so effective.

Feelings need to be shared with peers, and children are not their parents' peers. When one shares a negative emotion, there is an underlying need to be heard. The listener then responds with empathy, compassion, and assistance. The problem with sharing negative emotions with children is that they cannot respond without feeling overly responsible for the parent. It is fine for children to share feelings with each other or with their parents, but it is not okay for parents to share negative emotions with kids.

Children are hardwired to please their parents. If parents share negative emotions, then their child will feel responsible for comforting the parent. In a very real sense, the child begins to feel responsible to care for the parent. The child, who is the one who needs to be cared for, takes the responsibility to care for the parent. This reversal of roles is very unhealthy for the child. Girls will tend to lose touch with their own feelings and needs and care more for the parent.

Boys will tend to reject this responsibility and stop listening or caring.

It is very unhealthy for children to feel responsible for their parents' feelings.

When a parent is upset after a fight and then gets upset with their children, there is no way the children can understand they are not responsible for the parents' feelings. If the mother goes on to share her problems with her spouse, the children will feel even more responsible to solve the problem. It is hard enough for married adults to hear each other without feeling blamed or responsible. There is no way children can hear parents' negative emotions without taking on too much responsibility. Ultimately, this increased awareness of their parents' feelings will numb them to their own. As teenagers, they will eventually pull away and stop talking to their parents.

For example, telling a child, "I am worried that you will get hurt" or "I am sad that you didn't call" has the gradual effect of making a child feel manipulated and controlled by negative feelings. Instead, an adult should say, "I want you to be more careful" or "I want you to call me next time."

This is not only more effective, but it also teaches children not to make decisions based on negative emotions. The child cooperates not to protect the parent from the discomfort of feeling afraid, but because the parent has asked him to do something.

Children should never get the message that they are responsible for how a parent feels.

Parents can best help their children develop an increased awareness of feelings not by sharing their own feelings, but by empathizing, acknowledging, and listening. Using the five skills of positive parenting will automatically draw out your children's feelings.

ASKING CHILDREN HOW THEY FEEL

Just as a child should not feel responsible for a parent's feelings, children should not think that their emotions and wants put them in control. It is not wrong to ask a child how she feels or what she wants, but it should be done sparingly. If you are thinking about doing something and you ask your child how she feels about that, she may get the message that her feelings determine what you will decide. This gives too much attention to her feelings and wants and gives the wrong message that she is in control.

Directly asking children how they feel or what they want gives them too much power. Children need their parents to be in control, but also need to feel that their expressed resistance, feelings, wishes, and wants will be heard and considered.

Better than, "How do you feel about going to visit Uncle Robert?" say, "Let's get ready to visit Uncle Robert."

If they would rather go swimming, they will let you know how they feel.

Directly asking your children how they feel can sometimes have the opposite effect of what you want. A direction question puts too much pressure on children to know how they feel when they haven't yet developed self-awareness. Too many questions can awaken self-awareness too soon.

It is generally around the age of nine that children begin feeling embarrassed by things and experience greater modesty about their bodies. With this increased self-awareness, they are ready for more direct questions about feelings. Instead of asking a child how he feels, a parent can simply make an empathetic statement like, "I can see you are frustrated." At this point, the child's feelings and willingness to talk about feelings are stimulated.

The best way to teach awareness of feelings is to listen and to help identify feelings through empathy. Another way parents can create an understanding environment for feelings is by telling stories. Parents can successfully communicate that they too have feelings by telling stories of how they felt in reaction to some challenges in life growing up. This way, the child doesn't feel in any way responsible to help the parent or make things better.

Parents can successfully communicate that
they too have feelings by telling stories from
their past.

After a child talks about how afraid she is of taking a test, the parent might then disclose a sweet story from his childhood when he had to take a test and was afraid as well. Telling these stories should not only validate the child's feelings, but also give a reassuring message at the end.

WHAT YOU SUPPRESS, YOUR CHILDREN WILL EXPRESS

Even when a parent doesn't share feelings, children may still be affected by them. Some parents realize that it is inappropriate to share their feelings with their children, but they don't have effective ways of releasing them. As a result, particularly at stressful times, their unresolved emotions build up inside. Although they are holding them back, they can still affect their children.

When parents bottle up their own negative feelings, it will tend to intensify children's feelings. What parents suppress, their children will tend to express. It is often the most sensitive child in the family who takes on the unresolved emotional issues of the family. Therapists now commonly recognize that when a child has a problem, it is often linked to problems the parents are having.

It is often the most sensitive child in the family who takes on the unresolved emotional issues of the family.

If you are feeling anxious about deadlines, you may find that your children are always complaining about too much to do and not enough time. If you are feeling emotionally neglected or unsupported, one of your children is always complaining that he feels neglected. In these examples, what you are suppressing is being felt and taken in by one of your children.

It is as if they are sponges. If you are filled with love and empathy, they soak up your love to heal their own wounds and upsets. If you are filled with anxiety, depression, anger, sadness, fear, turmoil, resentment, or frustration, that is what they absorb. They literally take on your negative feelings and then act them out.

Children take on negative feelings and then act them out.

This explains why, on those days when you are extremely frustrated or overwhelmed, your children erupt with emotional turmoil or become extremely needy or demanding. When parents are not taking care of their own needs, the children will absorb that neediness and express it. Children act out at the most inconvenient times, because it is at those times that you are not giving yourself enough time as well.

Certainly, children have their own issues and emotions, but when they have to take on their parents' emotions as well they become overwhelmed and they explode in tantrums. Keep in mind that children will throw tantrums even if their parents are successfully dealing with their own emotions.

One way to determine that your child is acting out your feelings or their feelings is the resistance test. If you resist their feelings, then clearly they are expressing some of what you are resisting in yourself. If you are able to listen patiently with empathy, then clearly they are not acting out your unresolved feelings.

If you resist your children's feelings, then clearly they are expressing some of what you are resisting in yourself.

If you do feel resistance to their feelings it doesn't mean you are a bad parent. It is a clear sign that you need to take some time for yourself to nurture your own needs. It is fortunate that at those times when you can't be there for your children you can fall back on giving a time out. Regardless of what you are suppressing and the child is expressing, a time out will work to assist your child in dealing with what needs to be expressed.

Quite often, after giving a time out, parents feel better, too, because their child has expressed all their negative emotions. This is often why spanking and whipping children used to work so well to create temporary peace in the family. Not only did the child feel and express their pain, but the parents' suppressed feelings and pain was expressed as well. In this way everyone felt temporary relief.

THE BLACK SHEEP OF THE FAMILY

When parents have not learned to release negative emotions and instead suppress their inner feelings, at least one of their children will tend to take those feelings on, generally the most sensitive of the children. This child is often considered the black sheep of the family. Without an environment that accepts and nurtures negative emotions, these children either act out these feelings and become disruptive, or they turn the feelings inward and suffer low self-esteem. Quite often, both reactions occur.

These "black sheep" children are unable to get the nurturing and empathy they need, and the problems become worse. They will express the very feelings their parents are resisting and rejecting within themselves. Instead of feeling loved, understood, and embraced, they are resisted, resented, and rejected. They cannot get the love and support they need in order to process the strong emotions that they feel.

"Black sheep" children are unable to get the nurturing they need and feel something is wrong with them.

Often they don't know why they feel so upset and eventually conclude that something is wrong with them. The reality is that nothing is innately wrong with them. They behave inappropriately and get stuck in negative attitudes and feelings, because they are not getting the nurturing they need. These children can often get the support they need outside the family with others who are more understanding.

If one of your children tends to be the black sheep of the family, take extra time to listen to that child's feelings. Remember that every child is different and make sure that you don't ever compare children. This child needs lots of support from activities outside the family where he or she will not feel the pressure to take on and act out the unresolved problems and feelings of others.

MAKING NEGATIVE EMOTIONS OKAY

Making negative emotions okay is a completely different way of parenting. Never have adults and children had so many feelings. Never have we been so sensitive. The challenge of making feelings okay is great. It is not as if we have been raised by parents who knew how to handle and nurture the free expression of emotions. Yet with these new insights and skills of positive parenting, you will succeed.

As a result, your children will not be limited by life, but instead will be creative and capable of creating the life they want to lead. With a greater awareness of feelings, they will eventually know the truth of who they are and what they are here for in this world. They will still face challenges, sometimes even greater ones than their parents have, but they will have new and powerful resources for achieving their goals and making their dreams come true.

12

It's Okay to Want More

When children don't know what they want, they become vulnerable to the wants and wishes of others. They lose the opportunity to discover and develop who they are and, instead, become what others want them to be. In the absence of knowing what they want, they assume the wants of others and disconnect from their own power, passion, and direction. Without a clear awareness of their wants and needs, they are unable to recognize what is most important in life.

In the absence of knowing what they want, children assume the wants of others.

Too often children get the message that they are wrong, selfish or spoiled for wanting more or for getting upset when they don't get what they want. In the past, children were to be happy with crumbs, and that is what they got in life. They were to be seen but not heard and then, later in life, they were ignored and overlooked. They were not allowed to ask for more or even to want more.

The suppression of desire was an important parenting skill because parents didn't know how to deal with the negative feelings that would come up when a desire could not be fulfilled. The permission to want more gives children a power that parents in the past could not manage.

Today, with positive parenting skills for managing negative emotions, it is okay for children to want more. By wanting more, they can develop a stronger sense of who they are and what they are here to do in this world.

THE FEARS ABOUT DESIRE

It is often thought that giving children permission to want more will make them too demanding or difficult to manage. It is certainly much easier to parent a child who accommodates your every wish and desire, but this child doesn't get the opportunity to explore and develop his or her own sense of self, unique style, and direction in life. When children get the love and support they need to manage their feelings, giving them permission to want more does not make them demanding or difficult to manage. By wanting more and not getting, children learn the important skill of delayed gratification and self-discipline.

Some parents worry that it may make their child too selfish. This is true if parents always cave in to their children's wants and wishes. What spoils children is not getting what they want, but the power to manipulate others by wanting more and throwing tantrums to get it. Children become spoiled and selfish when parents deny their own wants in order to please their children.

Children become spoiled, not from wanting
more, but when parents stop wanting more
for themselves.

When parents seek to placate children by fulfilling their every desire as a way to avoid tantrums, then the children will become spoiled. To give children permission to want more, parents must be strong at those times when a child throws a tantrum and then give appropriate time outs. Given the opportunity to adjust their desires for more and accept the limits of life, children become even more appreciative of what they do have in the moment.

With regular time outs and good communication skills to assist their children in dealing with occasional strong feelings that come up, parents who give their children permission to think big and want more will raise confident, cooperative, and compassionate children.

By focusing on creating cooperation rather than blind obedience, positive parenting nurtures children's inner will and wish but, at the same time, maintains that the parents are in control. To create cooperation, it is not necessary to break children's will. Even if they want to stay up, children will go to bed according to their parents' will and wish. By applying the five skills of positive parenting, parents allow children to have their own wants and wishes, but reserves the final say in the end.

The problem with giving children permission to want more is that it does slow things down at times. Children are not always immediately compliant. They may want to do something else and will let you know. By taking this time to listen to and consider the merits of a child's will and wish, a parent nourishes a child's soul. When a child feels heard most of the time, then, at those times when a parent doesn't have the time, the child will be very accommodating.

Our soul can express itself through the will. When a child's will is not broken or ignored, it has a chance to breathe and grow. We are motivated in life by our will. Taking the time to nurture a child's will increases his or her bond with the parent and creates an overall willingness to cooperate.

All children are born with tremendous enthusiasm. This is the force of their will. When wanting more is accepted, this will is nurtured and can grow in harmony with their parents and others. But when it is not allowed to grow, children gradually lose that special spark we see in young children. The child loses that enthusiasm for life, loving, learning, and growing.

Nurturing a child's will sustains his
enthusiasm for life, loving, learning, and
growing.

By learning to feel their wants and to honor the wants of their parents, children develop the important skills of respect, sharing, cooperation, compromise, and negotiation. Without permission to want more and ask for it, children learn to sacrifice themselves for others. When children have permission to want more, they don't need to rebel as teenagers to find themselves.

THE VIRTUES OF GRATITUDE

Parents are too quick to teach the virtues of gratitude instead of giving their children permission to want more. "Be grateful for what you have" is too quick a reply to a child's desire for more. Many adults don't give themselves permission to want more in life because they are afraid of appearing ungrateful for what they have.

In the past, sacrifice has been a part of spirituality. To be good, holy, or spiritual meant to sacrifice in the name of

God. Sacrifice was a valid means to feel one's connection with God because it was a way people would begin to feel. To give up something for God forced one to feel more deeply. Today, we don't need to sacrifice to feel. We just need permission to want more and feelings will come up in abundance.

In the past, it was appropriate to sacrifice
for God, but today our challenge is to live
for God.

Our challenge is to create a life of abundance. We have so many resources at our fingertips today that were never available before. Your children can make their dreams come true. They can enjoy both inner and outer success. The basis of this success is permission to want more. Unless they have permission to want more, they will stop dreaming, and without a dream nothing will happen except what has happened before.

The secret of success, both inner and outer, is to appreciate what you have and to want more. A heart full of gratitude and love for what you have and a strong passion to achieve and have more is the secret recipe for our children's success in life. When children are able to manage the negative feelings that come up from not always getting what they want, then they will consistently come back to appreciating the love and support that they are given. With gratitude for their parents' love and support, children have the foundation to feel the desires of their own true self and not the misdirected desire that others may possess.

PERMISSION TO NEGOTIATE

When children are given permission to want more, it occasionally means more work for the parent. In return, children learn how to negotiate for more in life. Although negotiation takes more time, sometimes I am amazed by my children's power to motivate me to do things. I am proud of their strength and determination and resistance to blind submission of will.

When given the freedom to ask for what she wants, my daughter's inner power to get what she wants has a chance to blossom. She will not always take no for an answer. She is quick to negotiate and will often motivate me to give her what she wants. It is okay if she convinces you to change your position. This does not mean caving in to avoid the inevitable tantrum. There is a big difference between being manipulated by a whiny child and being motivated by a brilliant negotiator. Parents must maintain control throughout every negotiation by clearly setting limits on how long it can go on.

There is big difference between being
manipulated by a whiny child and being
motivated by a brilliant negotiator.

So many adults don't know how to ask for what they want, because they didn't have lots of practice as children. When they finally do ask for what they want, they still don't know how to negotiate. If they get a no, they either back off or feel resentful. So many adult problems would disappear if people had learned how to negotiate back and forth to get what they want. We have so many lawyers in this world simply because people are so inept in working out and negotiating their differences.

Children who have grown up negotiating to get more don't back off, nor do they become resentful. They know that no only means no if they don't come back with another good reason to comply with their request. They also know that no today doesn't mean no tomorrow. Negotiation requires a creativity and persistence that comes automatically when children are given permission to want more.

Children need permission to ask for more; otherwise, they will never know how much they can get. Even as adults we still have difficulty determining what and how much we can ask for so that we don't offend or appear too demanding or ungrateful. If adults have difficulty, then clearly we should not expect our children to know.

> Children need permission to ask for more;
> otherwise they will never know how much
> they can get.

When giving a child permission to ask for more, parents must understand and accept that sometimes a child will want way too much or seem very selfish. At these times, instead of judging or disapproving, parents need to give acceptance and understanding. A child cannot be expected always to know what is appropriate to ask for. It is a process of trial and error.

LEARNING TO SAY NO

Just because children have permission to ask for more doesn't mean that you will always acquiesce. Just as they learn to ask for more, a parent must practice being comfortable saying no. When parents can't say no, children very quickly will be unreasonable in their requests. They will keep wanting more and more until they reach the limit.

If parents are unable to set reasonable limits, children will ask for what is unreasonable. A child will push and demand for it until he or she gets a clear limit. When that limit is reached, the child will often need a time out to deal with the strong feelings of disappointment, rage, anger, sadness, and fear. The more children get what they want, the more upset they get when they finally don't get what they want.

If parents are unable to set reasonable limits,
children will ask for what is unreasonable.

When children negotiate for what they want, parents must set clear limits for how long the negotiation lasts. A parent only has so much time and inclination to continue negotiating. When you feel that you have heard enough and you are not willing to change, then it is time to say, "I understand you are disappointed, but now this negotiation is over."

If the child continues, the parent should just repeat this phrase and give a command to stop. The parent could say, "This negotiation is over. I want you to stop asking me."

If the child continues, he or she is out of control and needs to take a time out. After this process occurs a few times, the child will be very respectful of your request to end negotiations. Remember, when a parent shifts to command mode and repeats a command, no must mean no.

When a parent shifts to command mode and
repeats a command, no must mean no.

Most of the time, to end a negotiation, particularly with a young child, parents can use redirection. A mother might say, "I understand you are disappointed. I wish I could wave a beautiful magic wand and give you what you want, but I can't. Let's instead do this . . ."

There are two different situations when a parent is required to say no to a child. In the first situation, your child resists your request. For example, you want her to get ready to leave, but she wants to play more. In this example, you must clearly and effectively be able to say no to more playing and repeat your request. In the second situation, you are required to say no to her direct request. She wants you to play with her, but you have other plans. In both cases, the best way to answer is with confidence and brevity. Don't give a lot of reasons to justify saying no—just say no. If challenged, simply repeat the same response more firmly. These are some simple examples:

TEN WAYS TO SAY NO

1. No, right now I am busy.

2. No, I have other plans.

3. No, but maybe some other time.

4. No, right now I am doing something else.

5. No, this is what we are going to do.

6. No, right now I want you to . . .

7. No, but let's do this instead . . .

8. No, right now it is time to . . .

9. No, the plan is to . . .

10. No, right now I need to take a little alone time.

Besides giving children an opportunity to grow in their negotiation skills by giving clear "no" messages, children learn how to say no in their lives as well. Parents must not get upset with a child for asking for more. Permission to want more means the opportunity to ask. When parents clearly remember that they are not required to say yes, they can say no without feeling guilty. Besides making sure to take care of their own needs, parents who can say no comfortably provide a significant role model for their children. If a child continues to challenge a "no" message, all the parent has to do is simply say, "End of negotiation."

ASKING FOR MORE

One day, at about the age of six, my daughter Lauren asked me to walk with her to town to get a cookie. This was a little ritual we used to do together. One time, I told her no, and then she began to plead and negotiate. A neighbor happened to be there and immediately cut her off with a shaming message. The neighbor said, "Lauren, don't ask your father that. Can't you see he's busy? He can't say no if you keep asking."

Immediately, my three daughters, Shannon, Juliet, and Lauren, said in unison, "Oh, yes he can." It was a memorable moment. I was so proud. Each of my daughters clearly understood that she had the right to ask, and I had the right and power to say no in return.

When parents make huge sacrifices to accommodate their children's every request, this puts the burden of figuring out what is reasonable to ask for on the child. This is unhealthy. Eventually, the child will feel insecure about asking for anything. Instead, parents should give children permission to ask and give themselves permission to say no. In our house we often say, "If you don't ask, then you don't get, but when you ask, you don't always get."

If you don't ask, then you don't get, but when you ask, you don't always get.

Besides giving children permission to ask for more, parents need to also teach children how to ask. This is best done by modeling. When the parent makes requests in a respectful way, the child gradually learns how to ask.

MODELING HOW TO ASK

To teach children this important behavior, the most important technique is modeling. As we have explored in Chapter 3, instead of simply giving demands or commands, make sure that you make requests using "will," "would," "please," and "thank you." When children make demands or commands, instead of telling them not to be disrespectful, simply model a better way they could have expressed the request.

When a four year old says, "Daddy, give me that!" simply say in response, "Daddy, would you give me that? Sure, I would be happy to give you that." Then, simply give him what he asked for as if he had said those words.

This technique made parenting so much easier for me. When my children would become too demanding or sound disrespectful, instead of getting into a power struggle correcting them or trying to make them say what I wanted them to say, I simply modeled what I wanted them to say and then responded as if they had said it.

The only reason children don't express themselves in a more respectful way is that they haven't yet learned. We don't need to correct them; we just need to demonstrate what works. As parents, it is our job to teach them. As they continue to see that it feels good and it works, they follow suit.

If my daughter were angry and said, "Daddy, get out of my room," I would say, "Daddy, would you please leave my room? Certainly, I would be happy to." Then I would leave the room.

This clearly gives them the message of how to ask in a way that works. It would be a waste of time and energy to argue with my child, saying, "Don't tell me what to do. You ask me politely or I will not leave." This kind of approach just creates unnecessary resistance.

Children need to feel free to ask for what they want knowing that they will not be shamed. Even if it is not expressed perfectly, they should be respected in return. They also need to know that, just because they ask, it doesn't mean that they will get what they want. The way they ask should not be the reason a parent says no. When a child asks, she is always doing her best. If she fails, she is not bad; she just needs more modeling, or more nurturing, or a time out.

THE POWER OF ASKING

By giving your children permission to ask for more, you give them the gift of direction, purpose, and power in life. Too many women today feel powerless because they were never given permission to ask for more. They were taught to care more about what others needed and shamed for getting upset when they didn't get what they wanted or needed.

One of the most important skills a father or mother can teach a girl is how to ask for more. Most women did not learn this lesson as children. Instead of asking for more, they indirectly ask for more by giving more and hoping someone will give back to them without their having to ask. This inability to ask directly prevents them from getting what they want in life and in their relationships.

> Most women experience problems in their relationships, because they did not learn how to ask for more as children.

While girls need more permission to want more, boys need a particular kind of support when they don't get more. Quite often a boy will set his goals high, and parents will try to talk him out of his goals, because they want to protect him from being disappointed. They do not realize that, more important than achieving goals, is being able to cope with disappointment so that he can rise again to move toward his goals.

Just as girls need a lot of support in asking for what they want, boys need extra support to identify their feelings and move through them. For boys, this is best accomplished by asking for details of what happened while being *extremely* careful not to offer any advice or "help." Even too much empathy "to help him" can turn him off to talking about what happened.

Mothers often make the mistake of asking too many questions. When pushed to talk, many boys stop. When given suggestions on how to cope, boys particularly will back off. At times, when a boy already feels beaten, he doesn't need someone to make him feel worse by telling him how to solve the problem or what he did to contribute to the problem.

GIVING TOO MUCH

Whenever parents give too much, the children let them know. They become overly demanding and unappreciative of what they have. When you give something and they want more because it is not enough, it is generally a sign that you are giving too much.

When parents give too much, the solution is to pull back from making sacrifices for their child.

Let's look at an example. One day, my daughter Lauren wanted an ice cream bar. I was doing a series of errands but she really wanted one. Instead of doing what I needed to do, I agreed to get her one. Although I didn't realize it, I was giving too much. I was bothered by the interruption, but I did it anyway.

In the store, after waiting in a long line, she decided she didn't want the bar and wanted something else. She wanted me to go and find it. To do that I would lose my place in line and waste a lot of time. At that point, I realized that I was resenting her for being so demanding. She was clearly confused and asked if I was angry with her. I told her that I was just angry that this was taking so long.

Underneath it all, a part of me was angry with her but all she did was test the limit and ask for what she wanted. I was the adult—it was up to me to say what I could give and what I couldn't. That day I learned very clearly that it was up to me to determine what I could say yes to and what I couldn't. It was her job to keep asking until she reached a limit. It was not fair to let her push me over the limit and then blame her for being so demanding. By making sure that I didn't give too much in the future, and setting better boundaries, I would not resent doing things for her.

When my children occasionally behaved in a demanding manner in public, it was always a clear sign to me that I was giving too much. Whenever a parent gives too much to make a child happy, the result is that the child becomes overly demanding. Sometimes parents don't know if they are placating or pleasing their children too much. They are so happy to please them they don't realize they are caving in too often and giving them too much power.

CHILDREN WILL ALWAYS WANT MORE

When given permission to want more, children will always want more. At times it will seem like you can't make them happy. This is a healthy part of growing up. To realize their own inner ability to be happy, sometimes children need to experience that they can't get everything they want.

After not getting what they want in the outer world, they come back to feeling what they really need. As they feel their need for love, they suddenly begin to realize that they can be happy without getting everything they want. They don't have to have everything now. This is how children learn to delay gratification.

Children will also want more time and attention than is possible to give. Parents need to know that giving children permission to want more means that they will always want more. The lesson children need to learn is how to be happy now, even though they are not immediately getting what they want.

Delayed gratification is learning how to be
happy now even though you are not
immediately getting what you want.

This is why learning to manage negative feelings is so important. If a child wants more and doesn't always get it, he or she becomes very unhappy. Parents need to be careful not to try always to cheer up this child or to solve his problem. The child is like the little butterfly struggling to come out of the cocoon; it needs that struggle to strengthen its wings in order to fly freely.

To be happy in life, one of the greatest skills is delayed gratification. Learning to want more and yet be happy with what you have is a tremendous strength. This balance is developing each time your children are unable to have what they want and get upset. By assisting your children in feeling and then releasing the negative feelings, they experience again and again that they can be peaceful and happy in the moment, even though life isn't perfect and things are not the way they want them to be.

CHILDREN OF DIVORCED PARENTS

Children of divorced parents have a great need to process the loss of their parents' marriage. Deep inside, all children want mommy and daddy to love each other. Just as parents need to grieve the loss of a marriage, their children need to grieve that loss as well. Children often don't begin the grieving process until mommy or daddy has begun dating again. For this reason, once a parent has gone through the grieving process, he or she needs to begin dating again.

Children often don't begin the grieving process until they realize that mommy or daddy has begun dating again.

Sometimes single parents don't want to date because their children don't want them to. Their children get so needy and upset that single parents seek to avoid the tantrum by not going out. This attempt to placate the child will not only spoil the child, but also denies him or her an opportunity to grieve the ending of her parents' marriage.

In other cases, single parents don't start dating again because they want to compensate for the situation and give more to their children. The logic is clear: Since my child only has one active parent and a child needs two parents, then I must give more. This logic is correct, but the premise is wrong. Children will always want more than you can give. As a parent, you can only give what you can give.

If you try to give more, then you will be sacrificing too much for your children. Remember children are from heaven. If you do your best, God will do the rest. You can only do what you can do. The secret is to be there for your children when they go through resistance to your doing what you need to do for you. Rather than protecting children from not getting what they want, help them deal with feelings that come up in response to not getting what they want.

**Remember children are from heaven; if you
do your best, God will do the rest.**

When a single mom goes out on a date, her child may complain so much that she resists going out. She doesn't realize that other children of married couples also complain when they go out. When children feel safe to want more, they will. It is up to the parent to recognize what she can give without sacrificing her personal life as well. By taking time for herself, she is then able to give her child the quality time and attention her child needs most.

THE LONGING OF THE HUMAN SPIRIT

The desire for more is the longing of the human spirit. When children can manage to desire more, but patiently accepts and appreciates what they have, they are equipped to handle all of life's greatest challenges. The people who succeed in life are those who persist. The only failures in life are those who give up and stop striving, dreaming, and wanting. When their minds and hearts are open and their will is strong, there is nothing that can stop our children.

Children who are raised with strong wills will not submit to the will of a tyrant, nor will they seek to crush the will of others by means of dominance. They will model in the world a new way of interacting. Cooperation will be their daily experience and so they will have developed skills to enlist the cooperation of others as well.

Children who have grown up with the permission to want more think big and plan big. They are confident in their ability to achieve more. Contained within their deepest desires is the confidence and intuitive knowledge of how to get what they want.

By giving children permission to want more, this creative and intuitive awareness is awakened, giving them an edge on life that few people in the past have experienced. With this inner confidence and direction, your children will be able to move ahead in life, purposefully and with passion. Your children will achieve and well exceed what parents of previous generations could only hope for their children.

13

It's Okay to Say No, but Mom and Dad Are the Bosses

The basis of positive parenting is freedom. Each of the five positive messages gives children greater freedom to develop their full inner potential. This new way of parenting gives children the strength to move ahead in life without giving up who they are. With this support, children grow up learning that it is okay to be different, it is okay to make mistakes, it is okay to express negative emotions, it is okay to want more, and, most important, it is okay to say no.

The ability to resist authority is at the basis of defining a true and positive sense of self.

On the surface, positive parenting may seem permissive, but it is actually more controlling. The techniques establish control without fear or guilt. Giving children permission to say no does not mean that parents cave in to their child's resistance. Instead, parents hear and consider the resistance. Rather than mindlessly obeying their parents, children are given the opportunity to choose to cooperate.

Letting children say no opens the door for them to express feelings and to discover what they want and then to negotiate. It does not mean you will always do what the child wants. Even though the child can say no, it does not mean the child will always get his or her way. What the child feels and wants will be heard, and this in itself often makes a child much more cooperative. More important, it allows a child to be cooperative without having to suppress his or her true self.

Letting a child say no does not mean the parent will do what the child wants.

There is a big difference between adjusting your wants and denying your wants. Adjusting your wants means shifting what you want to what your parents want. Denying means suppressing your wants and feelings and submitting to your parents' wants. Submission results in breaking the child's will.

Without a strong will, your children are easily influenced by negative trends in society or peer pressure from other teens who are out of their parents' control. When a person does not have a strong sense of self, he is easy prey for others to manipulate and abuse. He will even be attracted to abusive relationships and situations, because he feels so unworthy and afraid of asserting his own will. Without a strong will, it is hard for preteens and teenagers to stand up for what they believe and too easy to be swayed by peer pressure.

Adjusting one's will is called *cooperating*; denying by submission of one's will is being *obedient*. Positive parenting practices seek to create cooperative children, not obedient children. It is not healthy for children to follow their parents' will and wish mindlessly or heartlessly.

Giving children permission to feel and verbalize resistance when it occurs not only helps them develop a sense of self, but it also makes them more cooperative. Obedient children just follow orders; they do not think, feel, or contribute to the process. Cooperative children bring their full self to every interaction and are able to thrive.

When children have permission to resist, it actually gives the parent more control. Each time a child resists and then surrenders her will to her parents' will, the child is able to experience and actually feel that mom and dad are the bosses. The ability to feel her connection to her parents' control provides the basis for positive parenting.

When children have permission to resist,
parents actually gain more control.

This felt connection sustains in children a strong willingness to imitate their parents' behavior and to cooperate with their will, while also providing the freedom to discover who they are, make mistakes and self-correct, feel and release negative emotions, want more and adjust to what is possible, and negotiate getting more. The permission to say no or resist authority is actually what keeps children aware that they are being controlled. It provides an essential safety net of security that supports each stage of a child's development.

HOW PARENTS AFFECT THEIR CHILDREN

To be successful in life, an adult pulls from a variety of inner resources. These resources are love, wisdom, power, confidence, integrity, morality, creativity, intelligence, patience, and respect, to name a few. The sum of all these resources is a person's unique perspective or consciousness. An adult decision or reaction to a situation is based on the adult's consciousness. When children feel an inner connection to their parents, they are able to benefit from their parents' consciousness. When children feel connected, they are in a sense plugged into their parents, and the light of their parents' consciousness affects everything the child says and does.

Connected children automatically benefit
from their parents' consciousness.

This parental consciousness gives children the security and confidence to be themselves and the ability to self-correct after making a mistake. As long as children feel connected, they automatically self-correct without long lectures or the threat of punishment. With the benefit of their parents' consciousness, children will automatically self-correct through trial and error.

Just being in the presence of an adult gives children the extra consciousness to behave harmoniously and creatively. Children always learn most effectively in the presence or under the supervision of a parent or teacher. The more connected children are, the more they are able to benefit from that supervision.

COPING WITH NEGATIVE EMOTIONS

Children can express and then release negative emotions because of the reassuring presence of their parents' consciousness. Children, before the age of nine, cannot reason, but with the support of an empathetic parent they can benefit from their parent's ability to reason and then release negative emotions.

Crying in the arms of a loving parent automatically heals the pain of a frightened child.

Crying alone with no one listening or caring reinforces a sense of abandonment, and the fear is not released. Children live in an eternal now. Without the ability to reason, they are constantly misinterpreting reality.

Just because children can imitate and
communicate, parents mistakenly assume
that children can reason as well.

When someone is mean, children assume the person will always be mean. If someone is loved more, then that person will always be loved more. If on the news someone is robbed, then a child concludes that he could easily be next. He cannot comprehend that he is safer because his house has locks on its doors. That conclusion requires logical thinking. He can feel safer if mommy or daddy hears his fear and then reassures him.

When looking for a good school for my children, I remember a comment made by another parent, who said, "It doesn't really matter what kind of teachers or kids there are at school. A kid's got to learn sometime that it's a jungle out there. It's better they learn now what they are going to deal with rather than later." Although this might sound compassionate and streetwise, it is not.

Children should be protected from the negativity of the world as much as possible until they have the brain capacity to interpret that reality correctly. When the body is developing in the womb, it requires the protection and support of the mother's body. Likewise, for the next nine years, children need protection from the negativity of the world. You don't prepare your children for a bad experience by giving them a bad experience.

A child is like a little seed sprouting which needs protection from harsh weather until it has a chance to be stronger. Children need to be protected from a bad teacher, a rough crowd at school, evening news, etc. Loving parents and family, supportive friends, and teachers provide the ideal womb for a developing child.

THE DEVELOPMENT OF COGNITIVE ABILITIES

Cognitive abilities develop much later in children. It takes nine years for their brains to develop the capacity for the logical interpretation of reality. Ideally, children should be protected from the harsh realities and negativity of the world until about nine years old. It is not until age fourteen that a child can consistently think abstractly, understand or propose hypothetical situations, reason logically on their own, and look at issues from another's point of view. A few years earlier preteens may have the beginnings of these cognitive abilities, but they have not yet fully developed. Without these cognitive abilities, children experience the world very differently.

As adults, we have forgotten what it is like to view the world without these mental abilities. From children's perspective, the world is a big place that can be upsetting, confusing, and create a lot of anxiety. The world today is even more negative and invasive than in previous generations. With technological advances in communication, children are being bombarded with negative information and stimulation all the time.

Children are being bombarded with negative
information and stimulation more now than
at any time in history.

When a child is kidnapped, raped, or murdered in
another state or country, the story is in your face wherever
you turn. It is on TV, in the news, in magazines, on the radio
and the Internet. When this news is brought into the home,
from your children's perspective it is as if the tragedy happened next door and could easily happen to them. Too much
exposure to the abuse and misfortune in the news numbs
children's natural sensitivity and weakens their feeling of
connection to parental control.

Repeated exposure to violence and crime
falsely normalizes what is not normal or
natural to life.

Children in the past were never forced to face and deal
with so many painful and negative realities of the real world.
Even adults have difficulty dealing with too much news
about the real world. Adults at least have brain capacity to
interpret world events more correctly—children do not.
Whatever parents can do to protect their children from this
intrusion will assist their children in feeling safe, confident,
secure, and protected.

CHILDREN'S NEED FOR REASSURANCE

Before the development of logical thinking, children need
lots of reassurance that everything is okay. Without the abil-

ity to reason or apply logic, children form incorrect beliefs and conclusions. Here are some examples:

When children don't feel loved, they conclude that they will never be loved.

If something is lost, a child believes it may never be found or be replaced.

If children can't have a cookie now, they think they will never get a cookie.

This insight helps parents understand why children have such strong emotional reactions.

Children are willful, feeling beings without the benefit of a logical mind.

When parents go away, children may conclude that they will never come back. Reasons cannot reassure children, but listening can. An empathetic response to children's tender feelings communicates a reassuring message, even though they cannot yet reason on their own.

Reasons cannot reassure children, but listening can.

The parents know they will be back and that everything is okay. This knowing is conveyed directly from the consciousness of a parent when she calmly and lovingly listens and reassures the child that everything is okay. By feeling connected, the child gains the benefit of the parent's life experience and consciousness.

CHILDREN HAVE A DIFFERENT MEMORY

Until about the age of nine, children have a different kind of memory. They can remember words, thoughts, and concrete actions. Since they have not yet developed logical thinking, they live more in the moment. It is unrealistic to ask a child younger than nine to remember to bring his lunch box or to put something away. He can learn to do these behaviors by repeated guidance and repetition, but should not be expected just to remember because it makes sense.

A mother mistakenly explains, "If you forget your lunch, then you will go hungry at school." A child cannot comprehend this reason or even think reasonably about her future. The best a parent can do is simply ask, "Would you please get your lunch box?" or "Would you please put that away?"

To expect too much from a child just reinforces a feeling that the parent is out of control and that the child is bad for being resistant or inadequate in some way. Neither of these conclusions is correct. The child is simply not ready to remember things because they make sense or are reasonable.

It is wounding to a child when a parent gets frustrated and says, "How could you forget?" The truth is, the child didn't forget, because he couldn't remember in first place. If anyone forgot, it is the parent, not knowing what to expect from children younger than nine years old.

COPING WITH INCREASED WILL

When wanting more, strong-willed children can eventually accept what is possible and what is not, because their parents have already learned to accept. The child benefits from the parents' experience that you may not always get what you want right away, but, if you don't give up, you will eventually get what you need. When children experience the pain of loss, delay, or disappointment, but feel understood, they connect with the maturity or expanded consciousness of the parent who is listening.

When children feel understood, they automatically connect with the maturity of the parent who is listening.

Strong-willed children will throw tantrums, but they will also gradually become more cooperative. Resistant children come back to their inner willingness to cooperate with their parents, because resistance itself creates the friction necessary to increase their feeling of connection. Children need to resist their parents from time to time in order to feel their connection. When they feel their connection once again, they are suddenly open and receptive to their parents' leadership and guidance. This new insight changes the way we view children's negative behaviors or attitudes.

When children are unruly or uncooperative, they are not bad—they are just out of control. They do not need punishment or to feel in order to self-correct or become more disciplined. Instead, they just need to come back into control. Whose control? Their parents' control. When parents apply the five skills of positive parenting, their children are once again back in control and happy to accommodate and cooperate.

Children are never bad, they are just out of control.

With positive parenting, children are not just being controlled, they are being given the ability to feel that control.

This is why positive parenting was not discovered before. Children in previous generations were not yet born sensitive enough to feel their parents' control. Without the ability to feel, children would not respond to positive parenting. Today, because a shift has taken place in the collective consciousness, these skills work for all children and teens, even if they were not raised with them. Children and teens of all ages will begin to respond right away.

Positive parenting skills work because
children today have a greater ability to feel.

Other more common, permissive parenting approaches have failed because they were not complete. It is not enough simply to let your children be and do whatever they want. To give children greater freedom, parents must provide strong leadership. By learning to balance increasing freedom with greater control, the skills of positive parenting are successful.

BALANCING FREEDOM AND CONTROL

By nurturing the feeling connection between parent and child, positive parenting provides a balance of freedom and control. Children experience the freedom to be unique and different, but also feel the strong need to imitate and learn from the parent. The freedom to resist actually strengthens children's sense of self, while simultaneously connecting them to the parents' will and consciousness.

The permission to say no helps children to identify their own desires, but, ultimately it strengthens their deepest desire to cooperate and gain the love and support of the parent. Without feeling awareness of their connection to their parents, children quickly forget their primal desire to cooperate. By using the five skills of positive parenting, the connection between parent and child is reestablished and once again the child is cooperative.

**When children are unruly, they don't
need threats of punishment; they just need
to reconnect.**

The permission to say no and want more creates strong negative emotions when more is not achieved. The expression of these strong emotions not only gives children the opportunity to learn how to manage negative emotions, but also increases children's ability to look within or feel. By making it okay to express negative emotions, a feeling awareness is generated that is necessary to connect child to parent.

Feeling awareness helps children identify their inner needs. With greater feeling, children are more aware of the need for their parents' love and guidance. Automatically, their willingness to cooperate and learn from their parents is activated. Rather than being shamed or punished, these children automatically self-correct with the benefit of their parents' innate consciousness of what is right or wrong, good or bad, smart or stupid. Although children don't directly know what their parents know, they are able to benefit from their knowledge and other resources to self-correct and to make needed adjustments.

Each of these five messages of freedom is counterbalanced with respect. It is okay to resist, but clearly mom and dad are in charge. The five freedoms cannot work unless parents maintain strong leadership. To give children freedom without authority is a kind of permissive abuse. Children need their parents to be in control. Using the love-based skills of positive parenting makes this possible.

TWO PROBLEMS OF LOSING CONTROL

When children disconnect from their inner willingness to cooperate with their parents' control, two very significant problems arise. They either act out or internalize the inner pain and turmoil of being out of their parents' control. Although some children will do both or alternate back and forth, generally speaking boys act out and girls internalize. To the extent that children don't get the support they need through parental control, they will continue to display some of the symptoms of being out of control.

Generally speaking, boys act out and girls
internalize their inner pain and turmoil.

Boys particularly become unruly and resistant to authority and misbehave. Without the support of family and parents, they become overly dependent on peers for leadership. One bad apple *can* ruin the whole barrel: A good kid is easily influenced by others who are not so good. When preteens and teens are disconnected from their parents, they are at greater risk of being "ruined" by peers who are even more out of control.

When out of parental control, boys tend to lose their ability to maintain their focus. They become hyperactive, which leads to increasing individual or gang-related aggression, violence, substance abuse, nonrelational sex, meanness, and cruelty. As parental control weakens, grades go down and school and family-related activities diminish. Cut off from parental control, boys are unable to develop their true potential.

As parental control weakens, grades
go down and school and family-related
activities diminish.

Though girls may experience some of these same problems, they tend to internalize the inevitable confusion and pain of being cut off from parental support. Girls particularly lose their confidence and their self-esteem drops. Instead of looking to their parents for support, some turn to boys and inappropriately use their sexuality as a way to get attention and feel special.

While boys become hyperactive and unable to focus or discipline themselves, girls tend to focus too much on their inadequacies and shortcomings. This leads to negative self-talk, over-gossiping, weight issues, petty meanness to other girls, poor self-image, teen pregnancies, suicidal tendencies, involvement in abusive relationships, drug addiction, and depression. Disconnected from the support they need from parents, girls are unable to develop their true potential.

THE NINE-YEAR STAGES OF MATURITY

Children need to move through three nine-year stages of maturity to become healthy and successful adults. For the first nine years, children develop best through growing in trust while being completely dependent. In the second nine years (ages nine to eighteen) preteens and teens develop by learning to trust themselves and by becoming increasingly independent. In the third stage of maturing (ages eighteen to twenty-seven), the young adult develops by becoming autonomous.

During the first stage, parents' challenge is to be completely responsible for the child. In the second stage, the parents' challenge is to maintain a sense of control, but to give a preteen and teen increasing freedom and independence. The process of letting go of control is gradual. Children cannot learn to trust themselves unless we give them the opportunity to be more responsible. Preteens and teens need increasing freedom to develop a healthy sense of responsibility. Just as in the earlier stage, parents should not expect perfection; children make mistakes at all ages.

**Preteens and teens need increasing freedom
to develop a sense of responsibility.**

In the third stage (ages eighteen to twenty-seven) parents need to back off and let go of being responsible for their children. Parents still hold the important role of being supportive in any way they can. This support is primarily determined by what the child requests and believes she needs and not what the parent thinks the child needs. For example, it is fine to give advice, lodging, or money if she is wanting or asking for it.

A loving parent needs to stop worrying about their adult child and instead admire his efforts to succeed. To worry about an adult just gives the message that you believe something is wrong with him or you don't trust him. Well-meaning parents wanting to make up for their past mistakes make matters worse by offering unsolicited help after a child is eighteen.

THE DEVELOPMENT OF RESPONSIBILITY

During the first stage, children are completely dependent on the support of parents for direction. Without complete parental control, they are forced to grow up too quickly and miss certain aspects of development. Learning to trust and depend on others is the basis of gradually being able to trust ourselves.

You could never learn to walk a tightrope if you had to practice high in the air without a safety net. In the beginning, you learn to walk close to the ground. Then you raise the rope, but have a safety net below. Without the security of knowing you can fall, it is impossible to learn a new skill. Knowing that you can depend on others and you deserve their support is a strong foundation for eventually developing independence and autonomy.

In the first stage, unless parents give clear messages that they are in control, children automatically assume too much responsibility. Children do not have the brain development to reason or look at issues from another's point of view. When they are loved, they assume it is because they are lovable. They assume responsibility. They believe they make others love them. When children are unloved, they assume they are unlovable and that they make others not love them.

Children assume responsibility for whatever occurs to them or around them.

Children are egocentric. The world revolves around them. When good things happen, they assume they make good things happen. When bad things happen, they assume they are responsible for those things as well. Unable to reason or look at issues from another point of view, they mistakenly assume too much responsibility.

For example, when a parent is in a bad mood, a child cannot comprehend that other things may be responsible for that parent's bad mood. The child immediately assumes that he or she is responsible. This tendency to assume too much responsibility can be corrected if the parent takes responsibility for his bad mood.

Even if the parent is nice to the child, unless she is doing something to nurture her own bad mood, the child will feel responsible. If the parent gets upset with the child, it makes matters even worse. The child feels even more responsible and concludes he is bad and unworthy. When parents lash out at their children, they need eventually to come back and apologize. Otherwise, children develop the belief that they are somehow responsible whenever their parents yell, argue, or fight.

Children assume too much responsibility
whenever parents yell, argue, or fight.

When parents argue, they should always do it in another room; otherwise, their children assume too much guilt. Ideally, parents should have good communication skills so they don't have to fight. If they do fight, they should do it quietly and in another room or location. When children even witness mistreatment of others, they will assume responsibility. Besides needing their parents to control them, children need their parents to be in control of themselves.

UNDERSTANDING THE GENERATION LINE

Parents and children are a generation apart. This gap must always be respected. Parents are above the generation line and children are below. To be above the line means to be responsible and in control. To be below the line means to be dependent on and in the control of the parent above the line.

When parents are calm, cool, collected, loving, peaceful, understanding, respectful, compassionate, and cooperative, they are in control of themselves and above the line. When children are dependent on their parents for control, they are below the line. Below the line children get to be children. With parents above the line, they have an opportunity to benefit from their parents' resources and consciousness.

With parents on top, children have an opportunity to develop all the necessary skills of growing up. Children are born with the potential to respect others, cooperate, forgive, self-correct, share, love, persist, and adapt, but they need the guidance and support of someone who has already learned these skills. When parents are above the line and children are below the line, they automatically draw upon their parents' consciousness as a resource.

When parents behave in irresponsible ways, either to each other, the world, or to their children, then their children cannot draw on their support. When parents are no longer in control and behave like children, they move below the line. When parents move below the line, then children move up above the line and become more responsible.

When parents move below the generation line, children may have to grow up too soon.

This is why yelling and screaming at your kids can make them jump back into control. When parents yell and scream, spank, hit, or punish, it is often when the parent has lost control and is now behaving like a child out of control. When the parent moves below the generation line by behaving like a child, the child goes up above the line and temporarily behaves more like a responsible adult. Although the child stops what he was doing, he is now on his own without the support of a parent in control.

If mom or dad is going out of control, then a child is left on her own and suddenly goes into survival mode. The problem with this survival mode is that she is forced to grow up too soon. Without a parent, she has to become her own parent. In this process of growing up too quickly, children suppress aspects of who they are and of their development. When parents go out of control, children quickly go back into control, but not in a healthy manner. Instead of learning to manage negative emotions and demanding desires, they cope by suppressing them.

This is similar to what happens when children are neglected or abandoned. For example, when a teen leaves home at age twelve, certain parts of who he is never get a chance to develop. He may be able to function like an adult, but certain skills like healthy emotional management or the ability to be dependent on others and ask for help are not developed. He may be very loving, but has greater difficulty being intimate or making a lasting commitment. This doesn't mean he can't function in life—it just means that something will be missing.

To give our children a chance to develop fully, they need eighteen years to be below the generation line. They need to feel that their parents are in control. They need to feel that they can depend on us to be there for them and that they don't have to be there for us.

DIVORCE AND THE GENERATION LINE

When parents get divorced, it also affects the generation line. In the absence of one parent, a child will often rise up above the generation line to comfort and nurture the wounded parent. It takes two parents above the line to raise a family of children below the line. When one parent is absent because of separation, divorce, or simply neglect, then children below the line will tend to rise up and replace the missing parent.

In addition, when a parent mistakenly looks to a child for emotional support and nurturing, the child rises above the generation line. When a parent needs emotional support, she should instead look to another adult who is above the generation line. It is unhealthy to try to get from your children the support you should be getting from another adult.

When a parent needs emotional support, she should instead look to another adult who is above the generation line.

For this reason, single parents are encouraged to get a life after divorce. It is not healthy to look to your children to fill up your life. Looking primarily to your children may feel good, but your children don't get a chance to grow up. They will grow up too quickly. They will tend to assume too much responsibility in life and then suffer from the various symptoms of guilt and unworthiness.

These are some of the symptoms of assuming too much responsibility:

Sensitive children tend to feel like victims in life and powerless to get what they want.

Responsive children often become people pleasers who deny their own wishes to accommodate others.

Receptive children become submissive, passive, and uncreative in life.

Active children become high achievers but are often disrespectful, too controlling, or abusive to others.

When parents behave responsibly and stay above the generation line, children have an opportunity to develop in a healthy manner. They still make mistakes and throw tantrums, but they develop the necessary coping skills for eventually becoming adults. Adults need to get what they need from other adults if they are to nurture their children below the generation line.

This insight frees single parents from the guilt that will often keep them from dating or pursuing a romantic love life. Although your children may resist your going out, it is still very important to go out. Your children need clear messages that you have a life separate from them and that you need others to make you feel good and nurture your needs for companionship, friendship, communication, romance, and fun. Unless you are taking action to get what you need elsewhere, your children will feel overly responsible.

CONTROLLING YOUR PRETEENS AND TEENS

Some parents will attempt to control a teen's activities too much while others will give teens too much freedom. There is no set list of correct rules or limits for each age. It takes common sense and trial and error to determine how much freedom your child can handle and when. Setting limits about behavior is a matter strictly between the parent and child;

whether it's watching TV, time on the phone, outside school activities, language, food and diet, time for homework, curfew, dating, friends, sexual behavior, chores, finances, appearance, and manners.

Giving children increasing freedom is an organic process based on lots of communication between parents and children and with other parents and teachers regarding what they consider to be appropriate. Although preteens and teens are given much more freedom to decide what they are going to do, each of the five skills of positive parenting is still necessary to maintain control.

If a sixteen year old is out of control and fights you in a disrespectful manner, then a time out is still required. When teens talk back, they need a clear message that their behavior is unacceptable and that you want them to be more respectful. As children become more articulate, parents need to remember that they still should not back up their requests with the threat of punishment or explanations.

If your teen resists your requests, then shifting to rewarding or commanding still works best. Just give your message and give it time to sink in. Until your children leave home, they still need you to be the boss, but they don't need punishment. With each year of maturity, they do need greater and greater freedom.

Until your children leave home, they still need you to be the boss.

Teens, just like children, need supervision. Even if you are not physically present, just their knowing that you know where they are and what they are doing goes a long way to assisting the child in feeling connected to your control. This

control is often lost when children become more independent in school. Children come home and parents ask how
the day went, but children have little to say. Sometimes,
before they can talk about their day, they often just need
some time to forget what happened or talk with a friend on
the phone about what happened. To get your teens to talk, it
is essential to maintain a presence in that part of their lives.

Parents are often so busy that participating in school activities and staying connected can be very difficult. For most parents, their primary link to the school is their children, and their
children are not talking. Children need their parents to be connected to school activities, and then children will more easily
talk about it. Fortunately, there are new technological advances that make this connection easier.

USING THE INTERNET TO IMPROVE COMMUNICATION

To maintain a better connection with their children's school
life and activities, parents can now use the Internet. More and
more schools are using new free software to connect schools
with parents at home. On the Internet parents can now access
their children's current grades, daily academic focus, events,
projects, homework, and progress reports, as well as e-mail
teachers and even connect with other parents. A convenient
personal calendar of your child's school and after-school activities, and a family calendar, including the activities of all your
children is also included. If your school is not already using it,
then ask your school administrators to begin. This new system
is free and available now to parents and schools.

On the Internet, parents can now directly access
information about their children's school activities.

Within a few minutes of logging on to the Internet, parents can become instantly informed with what their child is doing at school. With the click of a mouse, parents can send an e-mail message or ask a question to every other parent. To find out more about this software, provided free by Achieve Communications, just go online to their Internet site at www.achieve.com.

This increased awareness facilitates much better communication between parent and child. When parents simply say, "How was your day?" they can't really expect much of a response. But when they ask an informed, directed question, a door opens for better communication. Instead of general questions, parents can be more specific. Here are some examples:

How much progress did you make on your science fair project?

How did baseball practice go today?

How did Jessica treat you today?

How did your teacher respond to your oral report today?

I am really proud of the A you got on your math pop quiz.

Children can and will talk about their school life if parents have a greater awareness of what is going on at school. The more parents participate in school activities and are aware of what their kids are currently studying and who they are spending time with, the more their children will open up and talk to them. An ongoing dialogue about what is going on in our children's lives and at school is essential for staying in control and having influence over them. As parents become less influential, children become more vulnerable to the influence of other children.

An ongoing dialogue about what is going on at school is essential to having influence over your children.

Using www.achieve.com helps you connect with other parents and get an accurate reading on what they are doing with their kids. In raising my own children, it was almost impossible to get parents to meet regularly to talk about what their kids were doing and what they allowed their kids to do.

GETTING SUPPORT FROM OTHER PARENTS

As long as parents are not talking to each other, kids will have too much power. For example, they will come home saying they are the only kid who has to be home by 11 P.M., and you feel pressured to change the curfew. The reality is usually that only one parent allows a curfew of eleven while all the other parents have set the curfew at ten. Parents don't know this unless they are in communication with one another. Talking with other parents helps you figure out appropriate limits for your children and helps parents maintain control.

Start a monthly support group, read from *Children Are from Heaven*, and discuss the current issues you are having with your children. Talking about your issues and hearing what others are going through as well not only brings greater clarity, but also makes parenting easier. It becomes unnecessarily difficult when you feel alone. Parents need support, just as kids do. When you get the support of other parents, your children will get the support they need.

When you feel the support of other parents, it becomes much easier to maintain strong control and leadership while also giving your children permission to resist. With the support of others who share your commitment and insight, you will most effectively put the five positive messages into practice.

14

Putting the Five Messages into Practice

Deep inside all children, there is a button. When pushed, this button reminds them that they are willing to cooperate and they want to please their parents. The five messages and the five skills of positive parenting are primarily focused on that button. By learning to push that button again and again, parents gain the control they need to lead their children. Children at all ages primarily learn by cooperating and imitating. Unless parents keep pushing that magic button, children are restricted in their ability to learn and grow.

Children at all ages primarily learn by
cooperating and imitating.

Positive parenting even works with children or teenagers who were not raised with these five positive parenting messages. It is never too late to be a great parent and to inspire cooperation from your children. No matter when you start, by applying the five messages of positive parenting, you will hold the power to improve communication, create cooperation, and bring out of your children the best that they can be.

MOTHERS AND DAUGHTERS

The most strained or difficult relationship is often between mothers and their adolescent daughters, because often the mother is still trying to control everything in that child's life. To add fuel to this fire, daughters may resist their mothers even more because they were too accommodating when younger.

To develop a sense of self, adolescent girls feel a greater need to fight, defy, or rebel against their mother's control. When this is the case, using the five skills of positive parenting will teach the mother to maintain control in a healthier way, without having to raise her voice, make demands, express feelings, or use the threat of punishment. Mothers unknowingly create resistance by smothering children with reasons to act or behave differently.

FATHERS AND DAUGHTERS

Fathers often alienate their daughters by giving solutions and not asking enough questions. Men commonly don't understand the female need to talk and share, even when their daughters are not looking for advice or help. Men mistakenly assume that their job is always to fix things, when much of the time a little girl, teenager, or woman will just want to talk and be heard.

Since fathers tend to be more concerned with providing for the family, they are often less involved with the day-to-day details of raising the children. Often this gives little girls the message that their father just doesn't care.

The truth is, a father cares about his daughter's well being, and that is one of the reasons why he works so hard. At the same time, since he is not involved with the details, he doesn't care much about the little things. He is concerned about her general well being, but whether she wears jeans and a ponytail or a skirt with matching barrettes is not that important to him.

This is unfortunate, because when he doesn't care about the details of her life, she gets the message that he doesn't care about her. To bond with his daughter, a father needs to put in time asking informed questions and to practice listening without always offering advice.

MOTHERS AND SONS

Mothers often lose the respect of their boys by giving too many orders and then caving in when a boy is unwilling to cooperate. Mothers often complain that their boys will not listen to them. This is usually because they are giving too much advice and direction.

Boys generally need more independence and room to experiment than girls do. They have a greater need to prove what they can do on their own. Too much help is interpreted as a lack of trust, and eventually a boy disconnects and stops listening.

Boys generally need more independence and
room to experiment than girls do.

Mothers tend to use upset emotions or long lectures to control boys. This common mistake will quickly cause a boy to disconnect from her, and she will have even less control. When a boy is resistant or uncooperative in response to his mother's request or command, she must be prepared to face his tantrum and put him in a time out. If she simply gives up or waits for dad to come home, she gives up control.

How a husband treats his wife also makes a big difference in the way a son respects his mother. For example, if the mother has made dinner, and the father doesn't respect the mother's will and respond right away when the mother calls out that it is dinnertime, the children learn that they don't have to either.

Children are always watching and imitating. When the father doesn't respond to the mother's requests, it is a clear message that the boys don't have to listen either. This is another good reason why it is important to keep your complaints about your spouse private. When mom complains about dad's disrespect in front of the kids, she is unknowingly teaching her children that they don't have to respect her either.

FATHERS AND SONS

Fathers best bond with their sons through action. By doing things together, a son gets a chance to experience his father's appreciation, admiration, and assistance. While sons and fathers may not need to talk as much, they do need to bond. Fathers must be careful not to be too critical with their sons or become frustrated with their sons' deficiencies. A son needs a clear message from his dad that he is normal and acceptable just the way he is.

Fathers must be careful not to demand
too much from their sons.

Boys are naturally more goal oriented. When a boy fails, he needs to know that he can come to his father for understanding, comfort, and appreciation that he did his best. Fathers must hold back from always pointing out how the child could have done better. Fathers need to remember that all children are different and learn at different rates. Their job is to find children's strengths and acknowledge them with praise, pride, and admiration.

As fathers learn to control their sons without the need of punishment, a whole new door opens up for a son to bond with his father. No longer does he have to risk punishment from his father for making mistakes. He is free to go to his father for advice and leadership during all his years at home.

TEENS SECRETLY APPRECIATE LIMITS

Being the boss is different from being your child's best friend. Teens don't always like what you finally decide, but they will respect and accept it, *and* they will secretly appreciate you for it. For teenagers, peer pressure is strong. They can more easily resist temptations by blaming their parents. Although teens can't honestly say they will be punished by parents who use positive parenting skills, they can say they'll get into a lot of trouble with their parents.

Teenagers can more easily resist peer pressure
by blaming their parents.

Just because positive parenting doesn't punish kids, children are not free to misbehave without consequences. For teens to gain more freedoms, they have to earn trust. If they prove to be incapable of respecting the limits of their new freedom, then less freedom is given. A parent may temporarily take away a freedom as an adjustment, but not as a punishment.

A parent may temporarily take away a freedom as
an adjustment but not as a punishment.

For example, Tom, a sixteen year old, was allowed to stay out till 1 A.M. on the weekends. Yet he would consistently come home around 2 A.M. He explained that he would just forget, and he didn't understand why it was such a big deal. His mother, Sarah, explained that at night there is less supervision and one needs to be more mature and responsible. If he could not remember to be home by one, then he was not mature or responsible enough to stay up that late.

Sarah told him that if he were to stay out till 1 A.M., he would have to call home around midnight. This act of responsibility would help him to remember when to come home. Tom continued to forget. After several attempts to solve the problem his mother finally realized that she had made a mistake by giving him too much freedom too soon.

Sarah then said to Tom, "I realize that I made a mistake in letting you stay out till 1 A.M. I know you are doing your best, but I just don't think you are ready. From now on you can stay out till midnight, and if you do that on time for a while, I will reconsider a 1 A.M. curfew."

In this way, a freedom was temporarily taken away until Tom earned the necessary trust. Taking away a privilege should only be used as a last resort. Teens should clearly get the message that you have tried a variety of ways to work with them and you have finally concluded that they are just not ready for the increased freedom or privilege.

Parents must be careful they are not making an adjustment to punish, threaten, or make their teen more cooperative.

These kind of adjustments should only occur because parents realize that they were giving too much freedom and that they need to go a little slower. A teen can then gradually earn their trust to have more privileges. Keep in mind that adjusting a teen's privileges should only be done after you have tried other approaches. Remember if you are punishing your teens, they will not openly come to you for support.

Positive parenting doesn't punish, but it does adjust privileges when it becomes necessary.

To be in control, parents need to know where their kids are, with whom they are spending time, what they are doing, and who is watching what they are doing. Yet teens often just don't know where they are going. They want to get out, be together, and do whatever. If they are old enough to drive, then just driving around is enough for them to be happy. A parent wants to know where they will be and teenagers may really not know. With this problem, as with others, creative solutions can be found by making new rules.

If a teen is given permission just to go out with her friends, the new rule may be for her to call in at 10 P.M. or even to wear a pager. If the teen forgets to call or wear the pager, then the new freedom just to go out will be withdrawn until that child can remember to wear a pager or call in at a certain time.

Although teens resist these rules another part of them is grateful that you are staying in control and responsible.

For the next month, the teen is then required to know in advance where she is going, but she is also given the opportunity to practice calling home at 10 P.M. or wearing a pager. Once she proves she can remember, then the freedom just to go out is given again.

As a teens earn trust, they should be given greater freedom.

When a curfew is extended, some parents have their kids call in at certain times to talk to them and to make sure they are okay. It is helpful for teens to know their parents may call and talk to them at anytime. This is another deterrent against taking drugs or getting into trouble.

WHAT TO DO WHEN YOUR CHILD TAKES DRUGS

If children are caught taking drugs, or you have reason to believe they may be taking drugs, but deny it, then drug testing should be considered. Random drug testing has been very effective to ensure that your kids don't take drugs. Any school counselor will teach you how to administer the test. The peer pressure to take drugs is so great that the possibility of being caught by the test is an added deterrent. Being able to say that their parents may test them gives teens more support to say no to their peers about drugs.

It is never appropriate to ground a child for days, weeks, or months. The longer they are grounded, the more they turn off to wanting to cooperate. Instead of being the parent, you become the enemy. When dealing with difficult behavior problems, parents need to listen more and get their kids to talk. Instead of telling children why they are wrong, ask more probing questions like:

Why do you think I don't want you to take drugs?

What do you think about that?

What is your experience of taking drugs?

What have you heard about the effects of drugs?

What do you think about it?

What do you think I could do to help you to not take drugs?

What more do you want from me?

Getting teens to talk helps them find out what they think. When kids get a chance to share their opinions, then they are more respecting of you. Even if a child's opinion is different from the parent's, a parent needs to be accepting, but still command a teen to do what the parent requires. At all ages, children will have different wants and, ultimately, when these are heard they are willing to follow your lead. They may not like it, but they will cooperate.

DEALING WITH DISRESPECTFUL LANGUAGE

One mother asked how long her sixteen year old should be grounded for speaking abusively to her. The mother wanted to know if two weeks was fair. I instructed her just to give a time out the next time it happened. Since her teenager had not been raised with time outs and could only be controlled by punishment, the mother thought her daughter would just laugh at her.

Since the child was not accustomed to positive parenting at all, I suggested that she should give her daughter more time for the first few time outs. Instead of sixteen minutes, I suggested two hours. Her mother still thought her daughter would just laugh. She said, "Two hours is nothing compared to what she deserves."

I reminded this mother that her daughter never deserved to be punished. As a parent, she felt that way only because she was raised that way and didn't really know another way to control and teach her daughter correct behavior and manners. Eventually, this mother was willing to give it a try, even though she was still convinced that her daughter would laugh at her.

I explained that the reason her daughter was so disrespectful was that she was out of control. To learn to express respectful behavior, her child just needed to come back into control. When given a time out, the child would have an opportunity to feel that she was being controlled.

To learn to express respectful behavior,
a teenager only needs to come back
into control.

The next time the mother got into a big argument with her daughter, instead of allowing it to escalate into something even more ugly, the mother wisely paused and simply said, "It is not okay for you to talk to me this way. I am your mother, and I want you to respect me. I want you to take a time out. I want you to go to your room for two hours. During that time, you can't come out. In addition, during that two hours you can't talk on the phone to your friends."

The mother was surprised when her daughter reacted with outrage. The daughter said, "How dare you tell me what to do. I will not take a time out. You can't tell me what to do. I hate you. You are a . . ."

The mother, unable to pick her daughter up and put her

in a time out, just continued to command her daughter. She repeated her command a few more times. The daughter eventually went to her room, kicking the walls and screaming profanities with every step. When the girl tried to engage her mother in more arguing, the mother just repeated the command, "I want you to take a time out for two hours. During that time you are not allowed to use your phone."

The mother was in a state of disbelief. She could not imagine that just a two-hour time out would bring up so much resistance. At the end of the two hours, her daughter came out and apologized for being so rude and mean to her mother. Using a time out worked immediately in their family.

Once again, this approach worked so well because, the parent just needed to establish control so that the child could to feel their bond and connection. A parent doesn't need to ground a child for days or weeks in order to establish control. When a child is grounded, a parent actually loses all control. For children today, punishments are counterproductive and actually weaken parents' influence and control over their children and teens.

PERMISSION TO SPEAK FREELY

Around the age of twelve, my daughter Lauren started swearing occasionally. Each time, I would calmly ask her to use more polite language. One day she began to resist me. She replied that I sometimes used swear words—why couldn't she? I explained that as an adult I knew when and where to use them and that as a child she didn't. She also didn't know when such language was appropriate and when it was not. Before she would be free to use swear words, she would first have to learn to control herself and hold back until she could find the appropriate time and place.

At first she was very resistant. She said everyone else swore at school, and she should be able to as well. Feeling the new freedom of a preteen, she challenged me.

She said, "I don't want to stop."

I countered by simply making my request again. I told her, "I understand all the other kids use swear words, and it is not polite."

LAUREN: I will not stop, and you can't make me stop.

ME: I know I can't control what you do when I am not with you. I can't stop you from swearing with your friends, but I can stop you from swearing in front of me. Around me, I want you to use polite words.

LAUREN: And what if I don't? What are you going to do about it?

ME: I will just ask you to stop and remind you that I don't want you to swear. I want you to use polite language.

LAUREN: And what if I don't?

ME: If you continue, then you will have to take a time out.

That was the end of the discussion. We both had an unfriendly attitude for the rest of the evening, but then it was forgotten. She was clearly testing the limits of the new power and freedom of being in the middle school and was just getting used to the part of herself that wanted to challenge me.

A few days later, she swore again. By this time, we had both had time to think about this new challenge. When she got in the car, she began swearing about someone she found frustrating. My response was the same, "Lauren, I don't want you to use that language around me."

She said, "Daddy, it's hard not to swear. All the kids do it. I feel like it builds up inside me, and I have to get it out. I don't know what to do."

I told her, "I've been thinking about this and think I have a good compromise. I just want you to do your best to be polite. If you feel the need to get it out sometimes, I will give you permission to swear when you need to but to make sure you learn to control your swearing you'll have to ask for permission. You know how in the movie *Star Trek* the crew asks the captain for permission to speak freely? If you ask me first, then I will determine if it is an appropriate time to speak so freely."

Since that time, this creative solution has worked wonderfully. When she feels she has to say something mean or not nice, she immediately whispers in my ear with a little smile, "Permission to speak freely?" If I agree, she then happily speaks out her swear words or mean comments. In this way, she learned to control her inner feelings and speak politely when that was required.

MAKING DECISIONS

Another area in which parents give up control is by letting their children make too many decisions. When children are given too much independence before the age of nine, they are easily wounded by making the wrong choices or decisions. When parents let them make their own decisions, if the outcome is not desirable children begin to doubt their ability to make decisions and become insecure. This insecurity can last a lifetime, causing an adult to be indecisive or unable to make a lasting commitment.

When parents are in control, until they are nine years old children should never be responsible for making choices or decisions. Certainly, they can voice their wants, wishes, needs, and feelings but the parent should make the decision. Even up to puberty, a parent should still make most of the decisions. When a child says, "Do I have to go?" the answer should usually be yes.

> Until they are nine years old, children
> should never be responsible for making
> choices or decisions.

When parents directly ask their children what they want or how they feel about something, even though the parents are making the final decision, children may get the impression that they are in control and not the parent. Positive parenting skills promote listening and considering children's feelings and wants, but they don't encourage parents to ask for them directly.

Certainly, some asking is fine, but it is better when children express their feelings on their own or in resistance to your control. Instead of asking, "How do you feel about not going to the park?" a wise parent says, "You look upset that you can't go to the park." By not asking, a parent does not give a child the message that his or her feelings control situations, and too much attention is not put on the child. Children are not ready for this kind of control until they are nine years old.

THE CYCLES OF SEVEN

During the first seven years, children are primarily dependent on the parents or primary caretaker to develop a sense of self. During the next seven years (ages seven to fourteen), children are still dependent on parents, but a shift takes place and they become more dependent on siblings, relatives, and friends to determine a positive sense of self. During the third cycle of seven years (ages fourteen to twenty-one), teenagers and young adults look to peers and others with similar goals or expertise in achieving their goals to help define and develop a sense of self.

The first stage is a time of getting what they need from their parents or primary caretakers. In the second stage, they develop a sense of self by interacting with others in a safe environment. Children's greatest need is to play and have fun. This is a time when parents should try to make things as fun, safe, and easy as possible. When children have learned how to get what they need in the first seven years and how to have a good time in the second stage, then they are ready to work hard and discipline themselves in the third stage.

Children's greatest need between seven and
fourteen is to play and have fun.

It is a mistake to push children too hard in the first fourteen years. This is the time when they are supposed to learn how to be happy. The ability to be happy is one of life's most important skills. Happiness doesn't come from the outer world, it comes from within. It is a skill. Happy people are happy, regardless of the outer circumstances.

Many parents push their children to grow up too soon because they want their children to be happy in life. They don't realize happiness is a skill learned during the second stage. No matter how successful children may become in life, if they haven't learned to be happy early on, they won't be happy.

Happiness is learned through play. From the age of

seven to fourteen, children should be encouraged to play and have a good time. With this basis, they will be ready to work hard in school to prepare for working hard in the world. Too much pressure to make good grades or do chores around the house can prevent children from developing the ability to be happy and enjoy their lives. When children experience that learning and chores are fun, they will not only be happier in life but they will enjoy their work and continue learning for the rest of their lives.

Too much pressure to make good grades or
do chores can prevent children from learning
to be happy.

In the third cycle of seven, from fourteen to twenty-one, teens have a much greater need to get the support of other teens. This is when peer pressure dramatically increases. If parents have not nurtured a strongly felt connection by means of good communication skills, their teens will turn to their peers for support and will be at risk of being influenced by the wrong element.

WHY TEENS REBEL

In stage three, it is natural for teens to seek out the support of other teens, but, in doing this, they don't have to stop feeling their need for parental and family support as well. When children are raised with positive parenting skills, teens don't need to rebel in order to develop a sense of self. At each stage of development, they have already experienced the freedom to be themselves. As a result, they have no need to rebel.

> Teens rebel if they were not given
> enough freedom and support to be
> themselves as children.

To resist unhealthy pressure from other teens, your teen needs to feel connected at home. This is accomplished not by increased control, but by applying the five skills of positive parenting. Teens need someone they can come to for understanding, acceptance, advice, and direction. They will only seek out their parents' support if parents know how to give them what they need.

> Teens seek out their parents' support
> if parents know how to give them what
> they need.

Many teens today are rebellious because parents have used fear-based parenting skills. As soon as parents stop using punishment and other fear-based parenting skills and begin adopting positive-parenting skills, the need to rebel goes away. Even positive-parenting skills will not work if parents continue to control too much. Parents of teens are still required to give their teens more and more freedom. If not enough freedom is given, then once again their teens may rebel. To decrease resistance in teens, parents must always balance freedom and control.

IMPROVING COMMUNICATION WITH TEENS

With teens, parents must be careful to not offer unsolicited advice. Teens have just developed their ability to think abstractly and form their own opinions. They now have the ability to consider another's point of view, but first they need to have someone to hear and consider their opinions. Even if they ask you what you think, don't answer before first asking them what they think.

Taking time to have conversations with teens on other topics besides what you want them to do will minimize their need to resist your control. At this stage, they need to argue and express their unique or different opinions. Talk with them about what they are studying in history and social studies and hear their opinions.

Teens need to assert a different point of view. Even when you don't agree with their point of view, you can at least appreciate their logic. You might say, "I would never have thought of that," or "Well, I don't agree, but that sure makes good sense," or "That's the good thing about America, everyone has the right to his own opinion."

Even when you don't agree with their point of view, you can at least appreciate their logic.

Give your teens the opportunity to experience your open-mindedness in another context besides the issue of how late they can stay out. They will learn that it is okay to disagree and have different opinions through your appreciation of their logic and opinions. This is an important experience. If you are open to their opinions, they will not be so demanding when it comes to expecting more freedom than they are ready for. Unless given the opportunity to disagree about current events, they will feel a need to disagree with you personally.

Give your teens the opportunity to experience
your open-mindedness in another context besides
the issue of how late they can stay out.

Resistant teens don't want to be told what to do. Before using the command skills of asserting leadership, parents first need to hear the logic of a teen's objections. A parent can then say, "I understand you think you should be able to get a tattoo. I hear that everyone else is doing it. I will consider what you are saying, but right now I want you to wait until you are eighteen to decide about getting a tattoo."

Teens have a much greater sense of justice and fairness than younger children. When parents behave like dictators, teens are sure to rebel. Listening and working together to decide about how much freedom a teen should be allowed will strengthen the bond between parents and teens.

Before giving a command, parents should first ask for cooperation, listen to the teen's resistance, and respect the teen's opinion. Then parents can express what they want, which might sound like this, "I understand you believe this is not fair. You want to spend time with your friends, and I want you to be here to see your cousins. I know you don't want to do this, but this is important to me. I want you to be there. I want you to be friendly and polite to them for two hours, and then you are free to go."

RESPECT YOUR TEEN'S OPINIONS

It is always best to move teenagers in the direction of forming and expressing their own opinions regarding why you want certain things, but not why you feel what you feel. It is not healthy to ask, "Do you know how that makes me feel?"

Expressing your feelings will just have the effect of making the child stop listening or feel guilty. Most kids today will hear a guilt trip and turn the other direction. Remember children up to age eighteen depend on parents; they are not responsible for how parents feel, though they can develop an understanding of why you want what you want. When a teen resists your requests, rather than lecture, get him or her to talk. Ask, "Why do you think I want you to do this?"

To maintain control parents should make sure that they don't expect their children to agree with them. Teenagers especially need the freedom to think differently and form their own opinions. This is an important stage of their development. If you don't demand obedience or agreement, they don't have to rebel. With positive parenting, it is okay to disagree, but always remember that, in the end, mom and dad are still the bosses.

Teenagers especially need the freedom to think differently and form their own opinions.

Parents need to remember that it is more important to keep their children talking to them than alienating them by giving too much advice or criticism. Parents need to be sensitive to feel when their children are really asking for advice and when they are testing them to see if they can still talk to them.

If your teen comes home from school telling stories about kids breaking the rules, behaving disrespectful, or engaging in inappropriate sex, parents have to be careful to exercise restraint and not immediately begin preaching, teaching, correcting, or threatening.

In each of the following examples, notice what your first reaction would be, and then reflect on another way you could respond that would ensure that your teen continues to talk with you:

- Harry was cheating today on his math test.

- Tina was cussing out her boyfriend right in front of everyone.

- Chris was skipping class and making out in the audio-visual room today.

- I hit Roger today when he kept pulling on my hair.

- I think Mr. Richards is really stupid and boring. He expects way too much from us.

- Susan was really tired today. She stayed out all last night getting stoned.

When teens make these or similar statements they are testing to see if they can really talk, or if you are going to begin preaching, teaching, controlling, or correcting them or their friends. Instead of reacting directly, parents first need to ask the teen what he or she thinks. Then the parent could ask, "What do you think I think about that?" Remember teens will keep talking to you if you keep listening to what they think.

Instead of responding directly, parents first
need to ask teens what they think.

If a parent immediately begins correcting a teen's thinking and behavior or starts calling other parents and teachers to address the problem, the teen will stop talking. Instead of reacting to solve the problem, parents need first to stop and hold back from giving advice. Keep listening, and try to remember some of the things you did as a teenager.

It is more important to keep the lines of communication

open than to do anything about problems. To maintain influence over your teens, they must feel their connection, and that occurs primarily when they feel heard by their parents.

After hearing about the events of the day, sometimes something may need to be done. Maybe a particular child is having inappropriate sex or speaking in a mean way and their parents should be warned. Parents must be careful to not just take action and do something about it. They should first ask their teen what they think should be done about it. By listening to what your teen thinks should be done, then they will be more receptive to what you think. If something needs to be done, the two of you can figure out together an appropriate action.

When the lines of communication are broken, your teen is at risk of being influenced by peers who are clearly out of control or not a healthy influence. If some teen is being mean to your teen, but your child doesn't want you to call their parents, in most cases it is best to respect your teen's wishes. The teen knows, and you should know, that if you violate her trust and use the information she shares in a way she feels is unfair or unsupportive of her, she will simply stop sharing.

SENDING YOUR TEEN AWAY

Sometimes a spoiled teenager needs more than time in his room. Sometimes time spent with supervision in a developing country, staying with a favorite aunt, uncle, or grandparent, or in the woods with a guide will help him to regain his true self and his need for someone else to be the boss. Getting away from the family on supervised activities that are challenging for a teen can dramatically improve his attitude.

By feeling out of control and depending on someone else, a teenager can come back to feeling her basic need for guidance and support. A button gets pushed again that awakens in her a need for her parents' love and her desire to cooperate and please them.

Having an after-school job outside of the home, taking private lessons, or being on a team are great opportunities for a teen to feel his need to be taught, directed, and supervised. Teens need the guidance of someone outside the family. If they are not getting the guidance of a boss at work, a teacher, or a coach they are more at risk of following a misdirected teen. To maintain control at home, make sure your teen is getting supervision and direction outside the home as well.

INSTEAD OF "DON'T" USE "I WANT"

Before children have developed logical thought, around the age of nine, it is counterproductive to use the word "don't." When you say "don't run," children will form an inner picture of themselves running. Instead of slowing down they have a greater urge to run. Children learn through pictures. When children picture something in their mind's eye, action is soon to follow. It is as if they don't even hear the word "don't."

When you say "don't," you actually create a greater urge in children to do the very thing you are asking them not to do.

Right now, try not to think about the color blue. By trying not to think about blue, you are forced to think about it. When you tell a child, "Don't hit your brother," he is now seeing himself hit his brother. Using the word "don't" actually makes it more difficult for the child to cooperate.

Telling children "Don't play with your food" creates in

their mind's eye a picture of themselves playing with their food and actually increases their impulse and desire to play with their food. When given the opportunity, this urge comes up later and they begin mashing things up with their food.

Quite often a parent will ask a child something like this, "I told you don't throw balls in the house. Why did you throw that ball in the house?" In this case, a child's honest answer is, "I don't know." Sometimes they really don't know why they threw the ball.

Many times children are not thinking,
they are just acting out the picture in their
mind's eye.

With this insight, parents can give up the word "don't." If they forget and a "don't" pops out, then a parent can easily rectify the message. By rephrasing their request or command in the positive, the intended picture will get created. If you happen to say "don't run," then adjust the "don't" and say, "I want you to slow down and walk."

ASKING YOUR CHILDREN WHAT THEY THINK

Around the age of nine, children begin to develop the capacity for logical thinking. It is appropriate for parents to begin asking their children what they think. If a child asks for ice cream in the afternoon, the parent can say: "What do you think? Is that a good idea?"

Besides asking children more about what they think they should do, parents can also begin explaining the reasons why they want their children to do things. Before age nine, children really can't understand logical thought. Saying "I want you to go to bed now" is best for kids up to age nine, and then a parent can say, "I want you to go to bed by nine o'clock so that you will be well rested in the morning."

These are some examples of requests for children nine years old and older:

Would you please be quiet? Right now, I want you to listen because then I can explain what we are going to do.

Would you please stop hitting your sister? I want you to use your words. Hitting hurts her and makes her not want to play with you.

Would you please help me? I want you to bring your plate over to the sink, because doing dishes is a big job and it is so much easier with your help.

Would you please clean up this mess? I want you put your toys away, because when you leave them out someone could trip on them. The room looks so much nicer when you put your things away.

Would you please straighten up your room? I want you to put your things away. When you put things away, then you know where to find them again.

If a child under the age of nine asks "why" he or she should do things, then it is fine to answer with some reasons, but not if the child is resistant. Parents should keep in mind that children younger than nine really don't have the ability to comprehend or put into practice logic or reason.

A parent should motivate the behavior of children under the age of nine without reasons to back up the request. When children resist direction, the only reason a child should cooperate is because he is the child, you are the boss, and you want him to cooperate. Remember that, deep inside every child, his prime directive is to cooperate with your will and wish.

THE CHALLENGE OF PARENTING

Parenting has always been a challenge, but positive parenting is an even greater challenge. Although it takes extra time and effort in the beginning to learn, it is well worth it. In the long run, not only does parenting become easier, but your children benefit as well. As your children move through their different stages of growth, you will be prepared at each turn.

As you apply the new skills of positive parenting, you will occasionally trip and fall, but everyone does. Soon you will feel confident and at peace, because you will know you are giving your children what they need. You cannot change their inner destiny or take away their unique problems, but you can give them the required parental support to face adversity and achieve increasing success.

As with any new skill, there is a learning curve. Before it gets easier, it becomes more difficult. As soon as you think you have it working, there is a setback, and you don't know what to do. When your approach doesn't seem to be working, or you don't know what to do, that is the time to review *Children Are from Heaven*. In a short time, you will rediscover what you forgot. By once again applying the five skills of positive parenting, you will be back on track.

Even if you do everything right, remember that children are not perfect. They need to make mistakes and experience setbacks. They need problems and challenges in life to forge their unique character and set of strengths. Although your support is greatly needed, they come into this world with what they need to learn their lessons and do what they are here to do.

Just as you can't expect your children to be perfect, don't expect yourself to be perfect either. Making mistakes is a part of growing up, and it is a part of successful parenting. Children cannot grow strong if everything is too easy. Children cannot accept their imperfections if they haven't had many opportunities to forgive their parents for their mistakes and imperfections.

THE GIFTS OF GREATNESS

By giving your children the freedom to discover and express their true selves, you give them the gifts of greatness. All great individuals, thinkers, artists, scientists, and leaders in history were able to say no to past conventions and to think creatively. They had dreams and were able to follow their dreams. When others opposed them or did not believe in them, they had the strength to believe in themselves. Greatness is always forged through opposition. Every success story is filled with examples of having to push ahead against others. By means of the process of saying no to others or resisting common ways of thinking and not blindly conforming, creativity and greatness can emerge.

Each of the five messages of positive parenting supports the development of a strong sense of self and contains a special gift of greatness. They are:

1. Permission to be different, which enables children to discover, appreciate, and develop their unique inner potential and purpose.

2. Permission to make mistakes, which enables children to self-correct, learn from their mistakes, and achieve greater success.

3. Permission to express negative emotions, which teaches children to manage their emotions and develop a feeling awareness that makes them more confident, compassionate, and cooperative.

4. Permission to want more, which helps children develop a healthy sense of what they deserve and the skill of delayed gratification. They are able to want more, and yet be happy with what they have.

5. Permission to say no, which enables children to exercise their will and to define a true and positive sense of self. This freedom strengthens children's mind, heart, and will and develops a greater awareness of what they want, feel, and think. This permission to resist authority is at the basis of all the positive-parenting skills.

I hope this practical parenting guide helps you be the best leader for your children. Being a parent is difficult, but as we all know it is the most rewarding job one can have. To make your job easier, seek out the support of other parents who are also using positive-parenting skills.

Let this guide help you on your journey. May your children grow up confident, cooperative, and compassionate. May they be successful in both the outer world and their inner world. May their material dreams come true, and may they always experience lasting love in their family and friendships.

JOHN GRAY, PH.D., is the author of eleven bestsellers, including the phenomenal number one bestseller *Men Are from Mars, Women Are from Venus,* which has sold more than ten million copies worldwide. An internationally recognized expert in the field of interpersonal communication and relationships, he has been conducting seminars in major cities for twenty-six years. He is a Certified Family Therapist (National Academy for Certified Family Therapists), a Consulting Editor of *The Family Journal,* a member of the Distinguished Advisory Board of the International Association of Marriage and Family Counselors, a Fellow and Diplomate of the American Board of Medical Psychotherapists and Psychodiagnosticians, and a memeber of the American Counseling Association. He lives in Northern California with his wife, Bonnie, and their three children.

If you like what you just read and want to learn more...

Please call our representatives at 1-888-MARSVENUS (1-888-627-7836) for information on the subjects listed below, or visit John Gray's Web site at:

www.marsvenus.com

MARS-VENUS SPEAKERS BUREAU

More than 500,000 individuals and couples around the world have already benefited from John Gray's relationship seminars. We invite and encourage you to share with John this safe, insightful and healing experience. Because of the popularity of his seminars and talks, Dr. Gray has developed programs for presentations by individuals he has personally trained. These seminars are available for both the general public as well as private corporate functions. Please call for current schedules and booking information.

MARS-VENUS WORKSHOPS

Mars-Venus Workshops are interactive classes based on the bestselling books *Men Are from Mars, Women Are from Venus* and *Mars and Venus on a Date* by Dr. John Gray. His personally trained instructors facilitate these workshops worldwide. While millions of people have improved their relationships by reading these books, taking a Mars-Venus Workshop will deepen your understanding of this material and permanently alter your instinctive behavior while you participate in a fun, interactive, non-confrontational, and "male-friendly" class. For those interested in presenting Mars-Venus Workshops in their own community, please refer to our Web site at **www.marsvenus.com** for the most current information.

MARS & VENUS COUNSELING CENTERS

In response to the thousands of requests we have received for licensed professionals that use the Mars/Venus principles in their practice, John Gray has established the Mars & Venus Counseling Centers and Counselor Training. Participants in this program have completed a rigorous study of John's work and have demonstrated a commitment to his valuable concepts. If you are interested in a referral to a counselor in your area call 1-888-627-7836. If you seek information about training as a Mars & Venus counselor or establishing a Mars & Venus Counseling Center, please call 1-888-627-7836. Please refer to our Web site at **www.marsvenus.com** for the most current information.

VIDEOS, AUDIOTAPES, AND BOOKS BY JOHN GRAY

For further explorations of the wonderful world of Mars and Venus, see the descriptions that follow and call us to place an order for additional information.

John Gray Seminars
775 E. Blithedale, #407
Mill Valley, CA 94941-1564
1-888-MARSVENUS (1-888-627-7836)

VIDEOS

JOHN GRAY TWO-PACK VHS VIDEOTAPE SERIES

In these five 2-pack VHS tape series, Dr. John Gray explains how differences between men and women—Martians and Venusians—can develop mutually fulfilling and loving relationships. Series includes:

MEN ARE FROM MARS, WOMEN ARE FROM VENUS

(2-Pack #1)

Tape #1:
Improving Communication
(60 mins.)

Tape #2:
How to Motivate the Opposite Sex
(56 mins.)

MARS AND VENUS IN THE BEDROOM

(2-Pack #2)

Tape #1:
Great Sex
(80 mins.)

Tape #2:
The Secrets of Passion
(47 mins.)

MARS AND VENUS TOGETHER FOREVER

Understanding the Cycles of Intimacy

(2-Pack #3)

Tape #1:
Men Are Like Rubber Bands
(45 mins.)

Tape #2:
Women Are Like Waves (62 mins.)

MARS AND VENUS ON A DATE

(2-Pack #4)

Tape #1:
Navigating the Five Stages of Dating
(57 mins.)

Tape #2:
The Secrets of Attraction (71 mins.)

MARS AND VENUS STARTING OVER

(2-Pack #5)

Tape #1:
Finding Love Again
(107 mins.)

Tape #2:
The Gift of Healing (105 mins.)

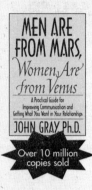
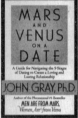
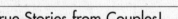

MARS AND VENUS STARTING OVER

A Guide to Recreating a Loving and Lasting Relationship

Whether newly single after a death, a divorce or other serious break-up, women and men will find comforting and empowering advice on overcoming loss and gaining the confidence to engage in new relationships.

Hardcover,
0-06-017598-2 $25.00

Trade paperback,
0-06-093027-6 $14.00

Two audiocassettes,
read by the author (abridged)
0-694-51976-6 $18.00

Keep Passion Alive!

MARS AND VENUS IN THE BEDROOM

A Guide to Lasting Romance and Passion

Hardcover
0-06-017212-6 $24.00

Trade paperback
0-06-092793-3 $13.00

Two audiocassettes (abridged)
1-55994-883-3 $18.00

Also available in Spanish:

**MARTE Y
VENUS EN
EL DORMITORIO**

Trade paperback
0-06-095180-X $11.00

Two audiocassettes (abridged)
0-694-51676-7 $18.00

The Keys to Making Love Last!

MARS AND VENUS TOGETHER FOREVER

A Practical Guide to Creating
Lasting Intimacy

Trade paperback
0-06-092661-9 $13.00

Mass market paperback
0-06-104457-1 $6.99

Also available in Spanish:

**MARTE Y VENUS
JUNTOS PARA
SIEMPRE**

Trade paperback
0-06-095236-9 $11.00

THE MARS AND VENUS AUDIO COLLECTION

Contains one of each cassette:
Men Are from Mars, Women Are from Venus; What Your Mother Couldn't Tell You and Your Father Didn't Know; and Mars and Venus in the Bedroom.

Three audiocassettes (abridged)
0-694-51589-2 $39.00

ALSO AVAILABLE:

**MEN, WOMEN AND
RELATIONSHIPS**
Making Peace with the Opposite Sex
Mass market paperback
0-06-101070-7 $6.99
One audiocassette (abridged)
0-694-51534-5 $12.00

**WHAT YOU FEEL,
YOU CAN HEAL**
A Guide for Enriching Relationships
Two audiocassettes (abridged)
0-694-51613-9 $18.00

✳ COMMUNICATION.
THE ANTI-DRUG.

A loving relationship cannot exist without communication. Research shows that kids believe they have valuable things to say. When parents ask them and listen genuinely, **it helps build self-esteem and confidence.** Also it demonstrates that you support their burgeoning independence as well as their ability to make intelligent decisions. The important thing to remember about drugs is that **it's not a five minute talk about sex. It's a dialogue.** As kids grow, they will need more information relevant to their exposure. In general, smoking marijuana is harmful. The younger a kid is, the more it may be. Research shows that people who smoke it before age 15 **are 7 times more likely to use other drugs.** It also shows that people who didn't smoke marijuana by age 21 were more likely to never smoke it. For more information, visit www.theantidrug.com or call 800.788.2800.

Communication is connection. During their teenage years, kids are exposed to an ever widening variety of people and influences. Know their friends as well as their friends' parents. Know your kids' routines and set curfews. Tell your kids that you love them. Praise them when they do well, no matter how small the accomplishment. Stay connected.